VEGETATION AND ENVIRONMENT

HANDBOOK

OF

VEGETATION SCIENCE

Editor in Chief

Reinhold Tüxen

1974

Dr. W. JUNK b.v. - PUBLISHERS - THE HAGUE

PART VI

VEGETATION AND ENVIRONMENT

edited by

B. R. Strain & W. D. Billings

1974

Dr. W. JUNK b.v. - PUBLISHERS - THE HAGUE

ISBN 90 6193 187 8
© Dr. W. Junk b.v. — Publishers, The Hague
Cover design: Charlotte van Zadelhoff
Printed in The Netherlands by Dijkstra Niemeyer b.v., Groningen

FOREWORD

During the International Botanical Congress in Edinburgh, 1964, Mrs. I. M. WEISBACH-JUNK of The Hague discussed a plan for preparation by her publishing company (Dr. W. Junk b.v.) of an international *Handbook of Vegetation Science*. She proposed a series that should give a comprehensive survey of the varied directions within this science, and their achievements to date as well as their objectives for the future. The challenge of such an enterprise, and its evident value for the further development of vegetation research, induced the undersigned after some consideration to accept the offer of the honorable but also burdensome task of General Editor.

The decision was encouraged by a well formulated and detailed outline for the Handbook worked out by the Dutch phytosociologists J. J. BARKMAN and V. WESTHOFF. A circle of scholars from numerous countries was invited by the Dr. Junk Publishing Company to The Hague in January 1966 to draw up a list of editors and contributors for the parts of the *Handbook*. The outline and list have served since for the organization of the *Handbook*, with no need for major change.

The different burdens of editors and authors have compelled quite different timings for completion of the individual sections. It was consequently decided by Dr. W. Junk b.v. that the parts of the *Handbook* would be published separately as they were ready.

The *Handbook* has as its purpose the presentation, through contributions of numerous collaborators from varied fields, of a comprehensive picture of modern vegetation science, including its development, its methods, its discoveries, and its goals, in approaches ranging from plant sociology as a descriptive science based on sharp observation of communities as concrete objects of study in the field (BRAUN-BLANQUET), through the inquiry into casually significant relationships that for many represents its principal goal, to abstract vegetation science that seeks a detached and generalized understanding based on quantitative relationships that may be represented in diagrams, equations, and models.

In this the *Handbook* should reveal the fundamental value of vegetation science for the other disciplines with which it is linked in many ways – not only such other areas of botany as plant geography and systematics, paleobotany and palynology, genetics and evolution, and others, but also soil science and the applied areas of forest, grassland, and wildlife management, water and fishery management and coastal protection, and interpretation of the land's potential for agricultural and other use by man. Not last in this is the decisive importance of vegetation science for nature and environmental conservation.

All these applied fields use vegetation, whether as an object of either harvest or preservation (or both), or as the essential context of the other phenomena of their direct concern, or as an indicator of relationships with which they deal. They all are in some way directed toward the care and maintenance of the green cover of our earth that they seek to use, wisely. For their purposes understanding of plant communities as functional systems and expressions of life-phenomena and the laws that govern them is valuable, or even quite indispensable. The plant cover of the earth is and must remain the basis for all other life.

So we hope to show that for men of many aptitudes and interests concern with the plant cover and its social units is rewarding – not merely in the sense of precedence and preference among individuals, but rather on the basis of the mutual concern of different fields and research purposes that complement and enrich one another. May our *Handbook* be a portrait – a composite portrait as painted by many scientists but harmonious as a whole – or one of evolution's achievements, the plant community as itself a harmoniously functioning, living system.

All our thanks go to the initiators, the editors, and the contributors, as well as Dr. W. Junk, b.v., Publishers, of the Hague. They together are making a lasting contribution to our science; may it also contribute in some measure to solve the grave problems for life on our earth.

Todenmann über Rinteln   R. TÜXEN

CONTENTS

1 Introduction, by B. R. STRAIN 3
2 Environment: Concept and Reality, by W. D. BILLINGS 9
3 The Ecological Niche and Vegetation Dynamics, by J. E. WUENSCHER 39
4 Description of Relationships Between Plants and Environment, by D. SCOTT 49
5 Allelopathy in the Environmental Complex, by C. H. MULLER 73
6 Correlation of Vegetation with Environment: A Test of the Continuum and Community-Type Hypotheses, by J. T. SCOTT 89
7 Plant Forms in Relation to Environment, by H. A. MOONEY 113
8 Modeling the Photosynthesis of Plant Stands, by C. E. MURPHY, T. R. SINCLAIR & K. R. KNOERR 125
9 Experimental Analysis of Ecosystems, by J. FRANK MC CORMICK, ARIEL E. LUGO & REBECCA R. SHARITZ . . . 151
Index Author 181
Index Subject 184

Authors and addresses

W. D. BILLINGS
  Department of Botany, Duke University, Durham, North Carolina, U.S.A.

K. R. KNOERR
  School of Forestry, Duke University, Durham, North Carolina, U.S.A.

A. E. LUGO
  Department of Botany, University of Florida, Gainesville, Florida, U.S.A.

J. F. MCCORMICK
  Department of Botany, University of North Carolina, Chapel Hill, N.C., U.S.A.

H. A. MOONEY
  Department of Biological Sciences, Stanford University, Stanford, California, U.S.A.

C. H. MULLER
  Department of Biological Sciences, University of California, Santa Barbara, California, U.S.A.

C. E. MURPHY, JR.
  School of Forestry, Duke University, Durham, North Carolina, U.S.A.

D. SCOTT
  Grasslands Division, Regional Station Department of Scientific and Industrial Research, Christchurch, New Zealand

J. T. SCOTT
  Department of Atmospheric Science, State University of New York, Albany, New York, U.S.A.

R. R. SHARITZ
  Savannah River Ecology Laboratory, Department of Botany, University of Georgia, Athenes, Georgia, U.S.A.

T. R. SINCLAIR
  School of Forestry, Duke University, Durham, North Carolina, U.S.A.

B. R. STRAIN
  Department of Botany, Duke University, Durham, North Carolina, U.S.A.

J. E. WUENSCHER
  School of Forestry, Duke University, Durham, North Carolina, U.S.A.

1 INTRODUCTION

I INTRODUCTION

In the past, the study of the interrelationships between vegetation and environment has usually been approached from the descriptive viewpoint. Synecologists generally have attempted to provide detailed quantitative descriptions of the vegetation but often have been satisfied with only cursory descriptions of the physical environment or of other biological features. Even now, when detailed environmental descriptions are obtained, most ecologists rely on statistical correlations and regression techniques to describe environment-vegetation relationships.

On the other hand autecologists, in an attempt to use an experimental approach, are generally forced to limit their attention to individuals or populations. The entire vegetation usually is considered too complex and varied to allow an experimental analysis. However, practical ecologists, *e.g.* range and timber managers, have often used experimental procedures to determine the effects of management alternatives. Unfortunately, such studies may have a limited usefulness beyond the immediate problem because of failure to interpret results in terms of basic ecosystem theory.

It is the purpose of this part to review the nature of vegetation-environment interactions and to make recommendations for their analyses. The part begins with a review of the concept of environment by W. D. BILLINGS. This paper is essentially an updating of his earlier paper (BILLINGS, 1952). The importance of the *holocoenotic* concept is reviewed and discussed in the light of new information and theory which has become available in the twenty-two years since the original paper was published.

Another concept of vegetation-environment interactions which is particularly important is that of the *ecological niche*. The complex idea of the ecological niche as a region of n-dimensional hyperspace is presented by JAMES E. WUENSCHER. His brief paper reviews pertinent concepts and recommends that vegetation scientists seriously consider making the niche concept particularly usable in a real sense. This recommendation is directed particularly to those autecologists who tend to consider only a limited number of niche parameters.

Very real practical problems are encountered, however, when one attempts to analyze the vegetation-environment complex experimentally and quantitatively in terms of holocoenotic or n-dimensional interactions. DAVID SCOTT discusses the complexities and recommends procedures for the recognition of relevant environmental variables. The use of causal diagrams and statistical procedures is recommended as a practical solution to the difficult problems posed by the multi-factorial approach recommended by the first two authors.

To complicate the analysis of the vegetation-environment complex even more, one has only to realize that the organisms and their biochemical products are environmental as well as vegetational components. Perhaps one of the best examples of this type of complexity is the phenomenon of allelopathy. C. H. MULLER discusses phytotoxicity, interference, competition, allelopathy and dominance as they effect the nature of the vegetation-environment complex.

As discussed in great detail in the preceding part edited by R. H. WHITTAKER, the classification of vegetation is complicated by theoretical considerations. If the true nature of the relationship between vegetation and environment were known, it would then be possible to test quantitatively the validity of the various classification schemes. JON T. SCOTT briefly reviews the vegetation-environment interaction from the classification viewpoint and makes a recommendation on how community environment might be employed in the classification of vegetation.

Dr. SCOTT reviews the nature of environmental gradients and presents suggestions for testing the plant association hypotheses. His paper calls for a large-scale cooperative study of the relation between vegetation and environment to improve our theoretical base in the classification of vegetation.

Textbooks in ecology and plant geography often refer to climatic and vegetation types. The general concept is that regions of the earth with similar climatic regimes (*e.g.* the Mediterranean Climate) have given rise to convergent plant form. The convergence of the succulent growth form in presumably unrelated plant families of Africa and the Western Hemisphere (*e.g.* Euphorbiaceae and Cactaceae) is a good example of the phenomenon. H. A. MOONEY considers the energetics of plant adaptation and proposes "For any given habitat condition there is a plant type which apparently represents the optimal form-behavioral strategy for carbon gain". Thus, the physical environment molds the vegetation into a form which could be predicted if the vegetation-environment interaction were better known. This information would presumably be useful for the further prediction of ecosystem structure and behavior.

As discussed earlier in the part, we have not yet found it possible to conduct total studies of vegetation and environment. And when the complexity of ecosystems is considered it is obvious why total system studies have rarely been attempted. An approach to this problem which has received some attention within the last few years, however, is to synthesize as much existing information as possible into a mathematical model of the interactions. An example of the approach is presented in the paper by MURPHY, SINCLAIR, & KNOERR.

As of now, however, no adequate mathematical model of the vegetation-environment interaction has been completed. Systems ecologists are still attempting to develop sub-process models. Once adequate models of components of the total interaction are available we anticipate that total system models will be derived.

The development and validation of whole system models would presumably be facilitated if procedures for the experimental analysis of intact systems could be developed. To accomplish this we desperately need experimental evidence from intact systems which exhibit all of the characteristics of natural ecosystems (*e.g.*, succession, stratification, zonation, niche differentiation, etc.).

The work of Drs. ROBERT PLATT & J. FRANK McCORMICK and their colleagues is suggested as an approach which should be used more frequently. The use of ecosystems from small depressions in granite rocks, as described by McCORMICK, LUGO, & SHARITZ in the final paper of this part, is presented as an example of experimental ecosystem analysis.

The papers presented in this brief part are representative of the work being conducted today on the environment-vegetation interaction. We have attempted to review the nature of the interaction and to make recommendations for future research into the structure and function of vegetation as an ecosystem component.

The authors invited to contribute to this part were asked to provide brief statements of the state of their science and to make recommendations for future work in their area. We believe that they succeeded admirably. Some editing of their manuscripts was required to create a uniformity in style. The editors accept responsibility for omissions or errors which may have been introduced in the rewriting.

2 ENVIRONMENT: CONCEPT AND REALITY

W. D. Billings

Contents

2.1	Introduction	9
2.2	Specific Environmental Components of An Ecological System	10
2.3	The Operation of a Specific Environment	16
2.3.1	Limiting Factors and Tolerance	18
2.3.2	The Holocoenotic Nature of an Ecological System	24

2 ENVIRONMENT: CONCEPT AND REALITY

2.1 Introduction

The concept "environment" may be viewed in either of two ways; "generalized" or "specific". Most people tend to consider environment from the generalized viewpoint. In this sense, one may speak of an "urban environment", a "desert environment", a "coastal environment", or the "lunar environment". Seen this way, environment exists without the necessity for referring to specific organisms or objects. Such an approach is useful in bringing to mind, in a few words, a general situation of any degree of complexity. Organisms, including man, may be present or absent in an environment conceptualized in this way; they may enter, stay, or leave. Organisms may change such a generalized environment from one type to another in the way that man "urbanizes" or "suburbanizes" a rural environment of the countryside around large cities.

A more rigorous usage of "environment" is necessary, however, if environment-organism or environment-object interactions are to be measured and modeled. This second concept I shall call "specific environment". It may be defined thus: "A specific environment is the whole of all influences, energy exchanges, and material exchanges through time between the universe and a particular living or non-living system". The living system may be an organelle, a cell, a whole organism, a population of genetically related organisms, or a community of organisms of different kinds. A living system with its specific environment is an "ecological system". At the community level, such an ecological system is an "ecosystem". A non-living system can be something as simple as a limestone pebble, or as complex as an unmanned planetary vehicle. A specific environment, then, is centered on some sort of living or non-living systems with which it forms a larger system. The interactions within this larger system, though complex, are subject to measurement once the system is defined and measurement techniques have been devised.

In view of the above statements, the question immediately arises as to the relationship of the "specific environment", as defined here, to the "operational environment" of MASON &

LANGENHEIM (1957). They define "operational environment" as the sum of those environmental phenomena which actually enter an operation of an organism from its conception to "now" as illustrated in Figure 1. Their "potential environment" constitutes all unused phenomena of the same kinds as are included in the operational environment. Potential environment lies between "now" and the death of the organism and becomes operational through time as long as the organism lives. I view "specific environment" in a somewhat broader (and perhaps less rigorous) sense as encompassing the "operational environment" including the history of the operation and the "potential environment" to the end of the environmental relation at death. This means that while both operational environment and specific environment are organism-centered, specific environment is a broader term which encompasses the whole life-span of the organism, group of organisms, or object. A specific environment may also include factors that operate indirectly on the organism or object through interactions in the environmental complex. In an operational environment, these indirect factors are excluded. Even though these indirect components act through other factors, in a realistic sense they must be included.

2.2 Specific Environmental Components of An Ecological System

As defined, the specific environment of an ecological system consists of all forms of energy and materials which modulate the rates of formation and operation of the gene action systems of an organism or a community in any way. Some environmental components operate continually, some in a cyclic fashion, and some only occasionally at irregular or random intervals. It is obvious that at any given time, many components are acting simultaneously but in different ways and at different rates on different metabolic processes and pathways. Many influences, processes, and materials are involved. While some components are so universal that all organisms are influenced by them, some are specific to a single individual of unique genotype, or to a population of a single taxon at a particular time. Gravity, although it varies in degree of force slightly from place to place is a near-universal component. An insect that pollinates the flowers of one tropical plant species at a given season of the year exemplifies a rather specific component. Even though one environmental component has overriding or dominant influence at a given time, the organism or community

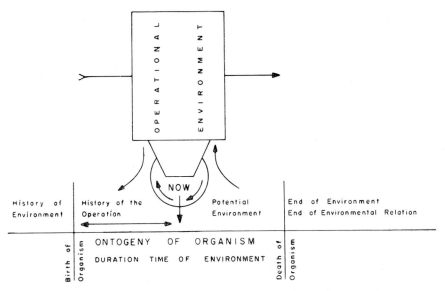

Figure 1. Diagrammatic summary of the theory of an operational environment (from MASON & LANGENHEIM, 1957, with permission).

remains within the aegis of the total environment. The total environment-organism complex is changed by the dominant component no matter how short a time involved nor how intricate the complex. The dominant component can be rather subtle and may not appear to be dominant even though its effect, given enough time, may be very apparent — or even drastic.

To simplify things, even though realizing that a specific environment-organism complex operates as a whole, it is possible to analyze such an environment (albeit, somewhat artificially) and to derive approximate quantitative values for most of its components and flux rates. However, each component or rate in such an analysis must be evaluated in the context of the rest of the environment-organism complex. The object of such an analysis must be the synthesis of a model of the system.

A specific environmental analysis must first identify the principal components (beyond the organism itself), their source levels, their characteristics, their flux rates, and cycles through time. Table 1 is a simplified and admittedly somewhat artificial scheme which attempts to organize these characteristics into a qualitative list which shows some relationships but no quantities or interactions. It succeeds Table 1 in BILLINGS (1952) but has a much different approach.

While it is attractive to think of a whole specific environment as having dimensions, it is extremely difficult if not impossible to define them. Of all environmental dimensions, time is the easiest to comprehend and measure. Beyond time, does a specific environment occupy space? Where does it begin and end? Probably,

TABLE 1

Components of a specific external environment of a terrestrial plant, vegetation, or biological community*

Source Domain	Physical Components	Biological Components	
Universal Space	Universal "space-time" gravity Meteorites and "space dust" Primary cosmic radiation particles Solar radiation electromagnetic particles		
Atmosphere	Secondary, tertiary, etc. cosmic radiation Sky, reflected, and thermal radiation Radioactivity, including fallout Atmospheric gases Pressure Wind Water (vapor, cloud, and precipitation) Heat and temperature Fire Pollutants	Plants Animals Microorganisms Man DNA	Space-Time Continuum
Lithosphere	Rock and soil particles Minerals Water Radioactivity Heat and temperature Gases Topography (indirect)	Plants Animals Microorganisms Man DNA	
Earth Mass	Gravity		

* For aquatic plants or community, interpose a liquid water source domain between atmosphere and lithosphere or substitute it for the lithosphere in deep water. This aquatic source domain will provide many of the same components included in the atmosphere and lithosphere – but not *all*.

there is no beginning and no end either spatially or temporally except in the birth and death of the organism. In this way, but not necessarily in others, "specific environment" is comparable to the "operational environment" of MASON & LANGENHEIM (1957) from birth to "now"; however, "specific environment" continues on until death. On the community level, a specific environment begins with the colonization of a new area, proceeds through a series of successional populations and communities, and ends in their eventual destruction by some geophysical or biological cataclysm. Throughout the time dimension, diurnal, annual, and longer cycles in many environmental parameters are characteristic as, for example, in temperature. On the other hand, some factors such as sedimentation and accumulation of heavy metals tend to be cumulative — at least within reasonable time limits.

Environmental gradients exist outward from the organism; but flow rates, influences, and interactions along these gradients are different and complexly intertwined within the whole. It is difficult to ascertain the source distance and direction of flow for anything except for radiation and, perhaps, water. The complexity of the environmental pathway directions, distances, and origins can be imagined if one thinks of the ground, air, radio, and radar traffic around a large international airport. Where are the dimensional limits?

Leaving the dimensional dilemma for now, the scheme in Table 1 should be examined more closely. Here, the problem of spatial dimensions is avoided by imagining a Riemannian space-time continuum curving outward from the center of the earth through the universe. This hyperspace can be divided arbitrarily for ecological purposes into 4-dimensional "domains" of an environmental component. The source domains for both primary cosmic radiation and solar radiation are in "universal space" — the former from many sources far beyond the solar system. Insofar as is known now, no biological components of a terrestrial environment originate in space, or are known to have reached here from such a source. But information on this possibility is, to say the least, incomplete. The future may tell a different story.

For terrestrial vegetation, there are at least four source domains: aquatic vegetation has a fifth (water) either interposed between soil and atmosphere for shallow lakes, or completely substituted for soil in deep bodies of fresh water or in the seas. Physical components move between the source domains themselves, and also between the source domains and the organism by radiation, conduction, mass flow, or free fall. Gravity operates in a more complex way than the table would indicate. It can be considered as the resultant of the interactions between the earth's mass and the mass of all other matter in the Minkowski-Einstein curved space-time universe. Since, on the surface of the earth, the mass of the earth is dominant, the resultant gravitational force is downward toward the earth mass source domain. Except for the solar source of energy and the effects of gravity, almost all of the environmental components of organisms in the biosphere originate in the source domains designated "atmosphere" and "lithosphere".

Since Table 1 is relatively explicit, it is not necessary to go into detail in regard to the physical components of an organism's or a community's environment. The important and necessary physical components are the short-wave solar and sky energy which is captured in photosynthesis and flows through the system, that energy which maintains the heat balance, and materials which cycle through the system. These latter include water, mineral nutrients such as nitrate, phosphorus, calcium, (and others), and the carbon dioxide and oxygen involved in photosynthesis and respiratory processes. Some of the materials are derived from the atmosphere; others from the lithosphere. Since almost all of these materials cycle through the atmosphere, biosphere, and lithosphere, the original source domain boundaries tend to become even more blurred and one needs to know the residence time and flow rates of these components in different parts of their cycles in different ecosystems.

Fire is an example of a physical component which is unique in being a short-term factor with long-term effects. Few plant species require it directly except, perhaps, in reproduction. Such an example is the opening of cones in closed cone pines: *Pinus serotina* on the coastal plain of North Carolina and *Pinus attenuata* in the Sierran foothills of California. However, the indirect effects of fire through its acute or chronic control of community composition and mineral availability are important and long-lasting. This is particularly true in tropical savannas, the temperate grasslands of the middles of continents, and in the coniferous forest ecosystems of western and northern North America.

No organism lives alone. Even among the widely distributed clumps of *Saxifraga* and *Dryas* in the polar desert of the Canadian Arctic, insects circulate carrying pollen. The intricate biological interactions and plant-to-plant competition in temperate ecosystems pale by comparison with the complex structure and operation of a tropical rainforest ecosystem and its thousands of species and great diversity of niches. In considering the interplay among biological components, it is necessary to think in terms of the whole ecosystem, its structure, energy flow, mineral cycling, and the basic food chain and food web structures. Such an ecosystem and the interactions between physical and biological components is diagrammed in Figure 2 in the form of an oversimplified two-pathway food chain community biomass pyramid. These two pathways are the "grazing" and "detrital" pathways of ODUM (1963) which, of course, are not completely separate in nature at any of the trophic levels.

A complicating factor among the biological components is man. Primitive man lived largely within natural ecosystems in spite of his use of fire and his carelessness with it. He also hunted, gathered, and gardened. But modern man with his technology dominates almost all ecosystems, including the earth itself, and makes any simple description of his effects out of the question. The changes wrought by man in extinction of species, alteration of ecosystems, encouragement of opportunistic weedy plants and animals, and creation of whole new modified or synthetic ecosystems make man the environmental component whose impacts, direct and indirect, are changing ecosystemic environments most rapidly, drastically, and at least until now, unpredictably.

The roles of microorganisms as environmental components from their key energy and mineral cycling position in detrital trophic level 2 to their parasitic position in disease at all trophic levels are recognized, if not always known. The inclusion of DNA as a biological environmental component, however, may appear to be contradictory. It has been customary for decades to separate environment and genetic material, and to bring them together only in the formation of the phenotype. Just as there is a tremendous number of physical and whole-organism environmental component and gradient complexes, the WATSON-CRICK (1953) model of DNA makes possible a genetic code capable of carrying an equally great amount and variety of genetic information. If the DNA carrying this information code can move into and through the environment, the implications on gene-environment relationships are rather important. In fact, it has been known since 1952 (LEDERBERG & ZINDER) that phage is able to carry genetic material from one bacterial cell to another by a process called transduction. From the subsequent work of KORNBERG and others that DNA itself is necessary for the operation of the enzyme (DNA polymerase) involved in DNA replication, it appears that there really is no boundary between "life" and "environment" at the molecular level. Even at the cellular and possibly at the organismic level, the discovery that bacterial virus DNA can affect metabolic rates and pathways by entering the genome of human cells (MERRIL, et al., 1971) indicates that some DNA is environmental material itself. The

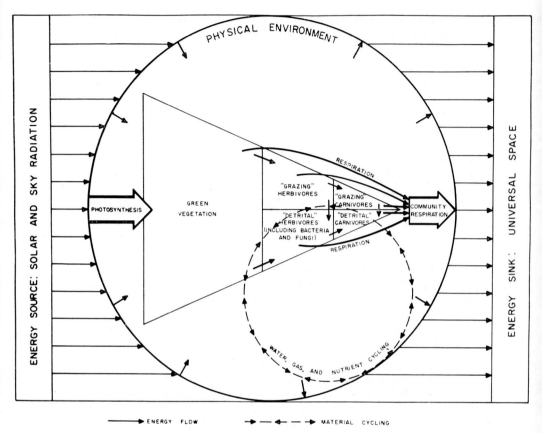

Figure 2. Diagrammatic scheme showing, in simplified form, the flow of energy and cycling of materials through the trophic levels of an ecosystem. The ecosystem consists of the "grazing" and "detrital" food chains as described by E. P. ODUM. Most energy modulates the physical environment; a small fraction is captured in photosynthesis by green vegetation, flows through the system, and is lost by respiration at all trophic levels. Man is a part of all trophic levels in both food chains above that of green vegetation.

implications of this knowledge reinforce the holistic relationship theory between "life" and "environment": they cannot be separated. And, further, it is possible that all organisms may be genetically related by DNA transmitted through the environment.

Concerning environmental components which control ecosystems, it is important and useful to think in terms of the rate of flow of useful energy, "power" (ODUM, 1971). All components of the physical environment, not only the flow of recognizable energy, may be thought of in terms of power, and are subject, of course, to the laws of thermodynamics and degradation and loss of useful energy. This approach is central to understanding the operation of environments and ecosystems. For detailed and imaginative treatment of the fundamental role of power in environmental and ecosystem management, one should read Odum's (1971) "Environment, Power, and Society".

2.3 The Operation of a Specific Environment

Like the "operational environment" of Mason & Langenheim, the "specific environment" of an organism starts to operate as soon as, if not before, the organism comes into being as a zygote. It is useful to think of four organismic levels upon which the total environment impinges. These are illustrated in the diagram as: (1) DNA, (2) RNA, (3) enzymes, and (4) the whole organism (see Figure 3). Through increasingly greater effects of environmental components, DNA results in RNA by transcription, RNA causes the production of enzymes by translation, and enzymes by initiating metabolic processes and regulating their rates produce whole organisms by growth. The increasingly wider arrows downward in the diagram indicate increasingly greater effects of the environment on each of these levels from least on DNA to the greatest effects on the whole individual organism itself. This latter set of effects may change as the organism grows to maturity. Environment also modulates the rates of transcription, translation, metabolism, and growth. Only at the whole organism level are there reciprocal effects of the organism on the environment — as in the shading caused by a tree which lowers the flux density of photosynthetically active radiation received at ground level. As the individual grows and environmental components change in degree with the seasons of the year, there is continuous readjustment in the plant-organism system regulated by positive and negative feedback from the whole organism to the subcellular levels of DNA, RNA, and enzymes.

2.3.1 LIMITING FACTORS AND TOLERANCE

Since the work of Liebig in the mid-nineteenth century, it has been known that environments can and do limit plant growth and reproduction. Such limitation can occur at any of the subcellular levels indicated in Figure 3, at the level of the whole plant, the local population, or the community. A limiting factor can be as simple as the lack of or excess of a single environmental component acting on a single organelle, process rate, or individual

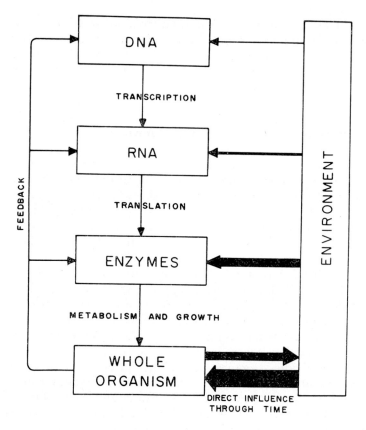

Figure 3. Diagram showing the main steps in the production of a whole organism by the interactions of environment, DNA, RNA, and enzymes.

plant. Or, the limitation may be much more general and widespread such as the lack of available water during a drought or the effects of a phosphorus-deficient soil on a whole biological community.

In uncontrolled natural communities, it is difficult to find which factors are limiting to each species at different times of the year unless the factor is so obvious as extreme drought in a normally mesic environment-vegetational complex. Within a single forest stand, the success of individuals of one species may be controlled by a biological pathogen while other species may be controlled by lack of sufficient nitrogen. With a uniform crop vegetation of a single variety of wheat, for example, it is much easier to find the limiting factor by experimentation in the field. If this factor is emended, growth will improve up to an optimum determined by the genome of the variety. Once this optimum is passed, the factor again becomes limiting. An example could be the optimum rate of net photosynthesis across a span of temperature; the rate decreases above and below that optimum. The situation is complicated, however, in that the characteristics of the temperature curve of photosynthesis depend not only on temperature but on light, carbon dioxide, vapor pressure, other environmental factors, and on the previous temperature history or temperature acclimatization of the plant itself. Once an optimum rate for a given plant or crop population is reached in regard to temperature, another factor such as light will take its place and become rate-limiting. The highest rate attainable will depend upon the genetic and phenotypic condition of the plant and finding just the right combination of necessary environmental factors. The search for such a combination is extremely difficult under natural conditions; it is the subject of experimentation under the controlled conditions of phytotrons and laboratory instrumentation.

As has been indicated earlier and is demonstrated in Figure 3, a limiting factor is not the product of environment alone but of the interactions between environment, DNA, RNA, enzymes, and single organisms. Some years ago, GOOD (1931) proposed his Theory of Tolerance which stated that "each and every plant species is able to exist and reproduce successfully only within a definite range of climatic conditions". As the last statement of this theory, GOOD added "The tolerance of any larger taxonomic unit is the sum of the tolerances of its constituent species". The corollary to this last statement is the tolerance of a species will, in turn, be made up of the tolerances of its constituent individuals (GOOD, 1964). In view of the discovery of the ecotype by TURESSON (1922) and the impact of modern genecology upon the species concept, the importance of a knowledge of tolerance ranges of individuals *within* a species cannot be overstated. MASON (1936) proposed that tolerances have a genetic basis, and both CAIN (1944) and GOOD (1964) accepted this in further elucidation of the theory. CAIN also made clear that different stages in the life history of an organism may have different tolerance ranges. Today, in light of the contributions from experimental genecology, these basic statements by GOOD, MASON, & CAIN seem very obvious, and yet they are important stepping stones toward an understanding of the relationships between environment and the distribution of species.

Advances in the last 30 to 40 years have been two-fold. First, considerably more is known now about how the genetic tolerance range of a plant interacts with environment to produce ecological adaptation by regulating physiological pathways and process rates. Secondly, much more information is available concerning the evolution of ecotypes and their tolerance ranges. HESLOP-HARRISON (1964) provides an excellent summary of advances in genecology up to that time. However, since 1964, knowledge of both of these aspects of tolerance has been accumulating at an accelerating rate.

Perhaps the most rapid progress has been made toward an understanding of the physiological aspects of ecotypic variation in tolerance. In view of the results of OLMSTED (1944), McMILLAN (1959), VAARTAJA (1959), MOONEY & BILLINGS (1961), and BILLINGS, et al. (1965), it is apparent that photoperiod length is very important in the origin of flower primordia, breaking of dormancy leading to flowering, and initiation of dormancy within many species of plants having a wide latitutinal distribution: ecotypes exist in this regard. Ecotypes also occur within many wide-ranging species in relation to light intensity. BJÖRKMAN & HOLMGREN (1963) found that ecotypes of *Solidago virgaurea* from shaded and exposed habitats have different light intensity requirements for maximum net photosynthesis. BJÖRKMAN (1968) since then has found that the ecotypic difference is due partly to the low carboxy-dismutase (carboxylating enzyme) activity intrinsic in the shade ecotype. But, this situation in *Solidago virgaurea* is not simple: light intensity during developmental growth is important also in this ecotypic differentiation in regard to photosynthesis. Moreover, there is an interaction in which the temperature of optimal photosynthesis increases with light intensity.

It appears that these results of BJÖRKMAN & HOLMGREN indicate a difficulty in separating light and temperature effects in photosynthesis and certain other processes. For example, SCOTT & BILLINGS (1964) found in alpine plants that at higher temperatures, more light is needed for photosynthetic compensation. And, PERRY (1962), while investigating day and night temperature effects on growth of *Acer rubrum* from different latitudes in eastern North America found that growth was optimum at temperature combinations typical of the latitude of the provenance; temperature ecotypes are present in this species of maple. But PERRY, too, found that it is difficult to separate temperature from light in this regard. Red maple from Vermont grew very poorly under low light intensities (1,000 ft-c, or less) while those from Florida grew quite well under relatively high temperatures and low light intensities. He concluded that temperature and low light intensity may counteract favorable photoperiodic conditions. While confirming the latitudinal photoperiodic ecotypes in regard to flowering of *Oxyria digyna* demonstrated by MOONEY & BILLINGS (1961), BILLINGS, *et al.* (1965) found that temperature was involved here, too, in that under continuous light, arctic populations come into flower much more quickly at low temperatures than do plants of this same species from alpine regions.

From the above information, it appears that ecotypes and ecoclines allow many wide-ranging plant species to be adapted genetically to the large range of light and temperature combinations in their environmental ranges. While there is less evidence in regard to water stress ecotypes, they do exist. MCKELL, *et al.* (1960) found that a western Mediterranean ecotype of *Dactylis glomerata* had a higher rate of water use and reached low leaf water potentials much sooner than an eastern Mediterranean ecotype adapted to longer periods of drought. Certainly, drought should be a strong selective force within a wide-ranging desert species. Indeed, AL-ANI, *et al.* (1972) have found that plants of desert populations of *Simmondsia chinensis* from the Sonoran Desert in southwestern North America when subjected to the same water stress in a common environment showed significantly less reduction in photosynthetic rates and water loss than plants from a coastal population of the same species from southern California. The photosynthetic rate of the coastal plants was reduced to 50 % when the leaf water potential reached only -20 bars; the photosynthetic rate in desert populations was not reduced until water potentials of about -36 bars. There appears, then to be "water stress-photosynthetic ecotypes" within *Simmondsia* which allows it to grow in deserts as well as coastal areas. Not all coastal-desert distributions show this water stress ecotypic differentiation, however. TOBIESSEN (1970) found that *Isomeris arborea*, in the same region as *Simmondsia* appears to have some photosynthetic and respiratory ecotypic characters typical of desert and coastal populations. However, the ability of *Isomeris* populations to withstand desert conditions is primarily morphological. A fast-growing deep root system enables plants of the desert ecotype to utilize flowing groundwater in canyons and under alluvial fans; but desert *Isomeris* plants cannot develop lower water potentials than can plants of the coastal ecotype of the species.

Turning to edaphic components of the environment, there is ample evidence of ecotypic adaptation in wide-ranging species to different geologic substrata and to different soil situations. Here, the adaptations are sometimes so marked that the ecotype has evolved into a species endemic to the substratum. One needs only to mention the many endemics and ecotypes limited to limestone, serpentine (RUNE, 1953; KRUCKEBERG, 1967), hydrothermmally altered rocks (BILLINGS, 1950), and rocks containing relatively high concentrations of heavy metals (ANTONOVICS, *et al.*, 1971). MASON (1946) and KRUCKEBERG (1969) have written perceptive essays on the restriction of narrow endemics and ecotypes to unusual rock types. Therefore, I shall not delve deeper into the subject here. It seems apparent that the most restricted species and ecotypes are edaphically limited. Climatic differences in environment are much less sharply defined than are the abrupt boundaries between rock types; the resultant climatic ecotypes and species thus tend to exist in continua or clines.

However, even edaphic endemics (whether species or ecotypes) are not controlled by substratum alone. TANSLEY (1917) showed long ago that competition between two closely related species of *Galium* (*G. saxatile* and *G. sylvestre*) is partially responsible for the restriction of the calcifuge species (*G. saxatile*) to acid substrata and the reciprocal restriction of its congener (*G. sylvestre*) to calcareous soils even though both will grow on either substratum when grown alone. Each of the species is slightly better adapted to one of the substrata than to the other; biological interaction makes the difference in nature. As a possible physiological mechanism for such a situation, CLARKSON (1965) has found that a calcifuge species of *Agrostis* (*A. setacea*) has a calcium transport system of lower capacity than three species closely related to it which do very well on calcareous soils. It is apparent that slight differences in mineral nutrient physiology may make the difference between success or failure of a species on different rocks — and that this success or failure is hastened by biological competition from close relatives which are slightly better adapted.

The adaptation to a specific environment is not just a matter of the evolution of populations of an ecotype or species in which tolerance is neatly pre-set by gene systems that produce a phenotype rigidly adapted to that environment. In out-breeding populations, there is the possibility of production of a great number of genotypes. Some of these genotypes produce phenotypes which are better suited to the environment than others. And, since most environments are anything but stable through time, some of these genotypes are more favored by environmental instability than others. Such genetic diversity within a population allows some species to adapt quickly to environmental changes while others cannot. Many weeds tend to fit the genetic diversity model while narrow endemics are more likely to have genetic uniformity within the population. On the other hand, a single genotype may be able to produce a great variety of phenotypes depending upon what environment it finds itself in at the moment. This "phenotypic plasticity" enables the individuals themselves to adapt to changing environments by growing faster or slower, by flowering or not flowering, or by adjusting metabolic rates and pathways in accordance with the whims of the operational environment.

Some of the plasticity of phenotypes is almost simultaneous with short-term environmental changes. The ability to close stomata under temporary water stress or the changes in photosynthetic rate as the sun becomes obscured by cloud are examples of such quick phenotypic response. Some phenotypic plasticity within a single genotype, however, depends a great deal on what the past environment of the phenotype has been. This effect has been called "acclimation", or perhaps in a better term, "acclimatization". I see no reason why such an effect should be confined to the individual's climatic history, even though in nature it is weather that does most of the changing. The addition of fertilizer to the individuals of a cloned pineapple crop will change the phenotypes of these individuals and their subsequent process rates and performance. Perhaps we need a new term: "environmentization". Whatever the environmental change, there is plenty of evidence that many phenotypes do adjust their process pathways, rates, and growth to past environment events, "acclimatization", and adapt for shorter or longer periods to temporal environmental events or cycles.

While acclimatization has been studied for some time in animal populations (see PROSSER, 1958), it has been only in the last decade or so that attention has been given to the phenomenon in plants. Most of this work has been on temperature acclimatization. SEMIKHATOVA (1960) reviewed the scanty literature on temperature acclimatization up to that time. Since then, considerable progress has been made in our knowledge of how plant individuals acclimatize in adapting to catastrophic or cyclic changes in climate Many of these investigations have been made on plants in the deserts and mountains of the American Southwest (MOONEY & WEST, 1964; STRAIN & CHASE, 1966; MOONEY & HARRISON, 1970) or on arctic and alpine plants (BILLINGS & MOONEY, 1968; BILLINGS, et al., 1971). MOONEY & HARRISON concluded that temperature conditioning in *Encelia californica*, a shrub of the semi-arid southern California coast, affects photosynthesis rates through a complex of causes including changes in stomatal and mesophyll resistances to CO_2 transfer, the degree of oxygen inhibition of photosynthesis, and probably some rather rapid biochemical changes. Certainly, these biochemical changes include the HILL reaction in photosynthesis (BILLINGS, et al., 1971) and the activity of ribulose-1,5-diphosphate carboxylase (CHABOT, et al., 1972). *Oxyria digyna*, an arctic-alpine species, was the subject of both of these latter investigations. In both HILL reaction and ribulose-1,5-diphosphate carboxylase activity, alpine populations were more responsive to the temperature of previous growth than were plants of arctic populations. Also low growth temperatures produced higher rates or activities than did high temperatures. These results are in agreement with those on the effects of growth temperature on net photosynthesis and dark respiration obtained by BILLINGS, et al. (1971) who found "ideal" homeostasis (CHRISTOPHERSON, 1967) of photosynthetic rates in alpine populations of *Oxyria* but only "partial" homeostasis in these rates in arctic plants of the same species. While *Oxyria digyna* consists of ecotypes adapted to low temperature growth, plants of such ecotypes are still subject to temperature acclimatization which regulates their photosynthetic and respiratory processes. Alpine ecotypes appear to be more capable of such acclimatization than are arctic ecotypes of the same species. This is not surprising, perhaps, in view of the changeability of alpine weather compared to that of the Arctic. In fact, BILLINGS, et al. (1971) conclude that acclimatization itself in this particular case seems to be ecotypic, and thus genetic. BRADSHAW (1964), had earlier suggested that phenotypic plasticity, *a priori*, must be genetically determined — and so it seems to be.

It has been shown by ANTONOVICS (1971) and ANTONOVICS, et al. (1971) that selection pressures resulting in edaphic ecotypes tolerant to heavy metals can be intense. Populations can differentiate into such ecotypes rather quickly in spite of gene flow between them. In fact, most heavy metal mine dumps in Britain are only between 50 and 100 years old; and yet, heavy metal tolerant ecotypes already exist on them. These ecotypes have been selected out from the seed rain of normal populations in adjacent meadows. SNAYDON (reported in BRADSHAW, et al., 1965) has the most startling data in this regard. He found zinc-tolerant ecotypes of *Festuca ovina* and *Agrostis canina* directly under galvanized fence that had been erected less than 30 years before — and this was several hundred miles from any zinc-mining area. TEERI (1972) found a similar situation in the High Arctic on Devon Island where snowbank and wet meadow ecotypes of *Saxifraga oppositifolia* are selected out from the seed rain of a nearby population which lives on a dry uplifted beach ridge. How long this process takes is not known but since the snowbank and wet meadow ecotypes do not reproduce from seed in that site, the selection would seem to be continuous. Summing up, it appears that ecotypic evolution in relation to environmental changes can be rather fast in certain species. Since environmental changes are occurring at accelerating rates, such evolution in regard to tolerance must keep pace or a species is doomed to extinction.

2.3.2 The Holocoenotic Nature of an Ecological System

The first conclusion of my 1952 paper was that the environment is holocoenotic. The statement was not original with me. However, at that time, I elaborated on the idea and attempted to reconcile the holocoenotic viewpoint with the limiting factor principle by the use of the "trigger factor" concept. The trigger factor principle is corollary to, and provides evidence for, holocoenotic relationships within an ecological system. Since that time, considerable evidence has accumulated to indicate that ecological systems are indeed holocoenotic; I know of no contrary evidence. In many of these cases, the trigger factor is known; in other examples, the visible environmental changes have appeared suddenly and without apparent reason. One can be sure that there is a reason but the reaction within the system is too far along before it is realized that change is occurring. The trigger factor for such an accelerating chain reaction within an ecological system may be identified from known facts. On the other hand, our knowledge of past environments is often too incomplete to do more than hazard a guess as to what event initially triggered the rapid changes observed now.

The concept of the universe as a holocoenotic complex is centuries old in philosophy and literature. Scientists were paying some serious attention to it by the nineteenth century. Darwin certainly considered the environmental screen in natural selection as a complex. Another great naturalist, John Muir observing the subalpine vegetation of the Sierra Nevada of California wrote in his journal on July 27, 1869: "When we try to pick out anything by itself, we find it hitched to everything else in the universe" (Muir, 1911). However, it was not until the twentieth century and the development of ecology that the holocoenotic concept was formalized.

In my 1952 paper, I ascribed the formalization of the holocoenotic principle to Allee & Park (1939). However, Friederichs (1927) had defined earlier what we now know as the "ecosystem" as the "holocoen" in which no one factor can change without affecting all others. Friederichs (1957, 1958) later expanded upon his earlier ideas concerning the holocoenotic principle. Allee & Park in 1939 said "To the working ecologist the environment is holocoenotic, that is, it is a unit composed of many

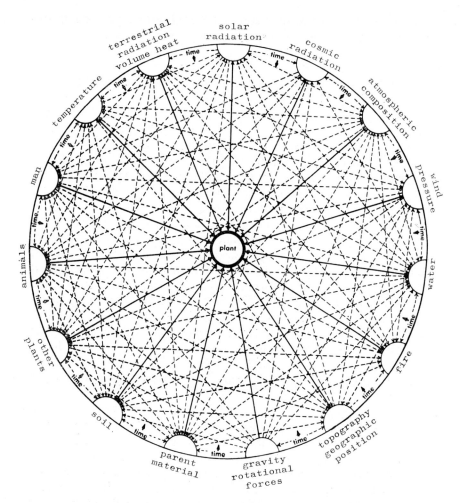

Figure 4. Diagrammatic representation of the interactions between environmental components and a plant in a holocoenotic environmental complex (from BILLINGS, 1952).

parts, as a rope is made of many strands". By the 1930's, then, the holocoenotic principle was already firmly established in ecology even though not always given its due. CAIN (1944), citing ALLEE & PARK, stated "The environment is holocoenotic" as one of his classic principles of plant geography. Since that time the principle has been firmly established.

To illustrate my concept of the intricate network of interactions in the holocoenotic environment of a plant, I included in the 1952 paper a theoretical diagram which is included here as Figure 4. This diagram was based on ideas and not on quantitative data. After twenty years, it still seems appropriate and workable as a model. Recently, DUNCAN PATTEN (1972) has used this model in a computer program to test the significance of some interactions between certain environmental factors and vegetational cover in non-forested ecosystems of Yellowstone National Park. Even though PATTEN did not test all possible interactions, his results, using different levels of significance, can be diagrammed in a fashion similar to the original model. The resemblance of PATTEN's real data in Figure 5 to the hypothetical model of precomputer days is remarkable.

Today, there are many examples of holocoenotic reactions within ecosystems which were not available to me in 1952. In some cases, the trigger factors are known; in many other cases, the reaction has been observed but how it was triggered remains unknown. Before presenting some of this evidence, it is appropriate to have a second look at the two main examples of holocoenotic effects mentioned in the 1952 paper. In both cases, the trigger factor was known.

The first case discussed in 1952 concerned the accidental introduction of the Mediterranean weedy annual grass, *Bromus tectorum*, into the *Artemisia-Poa-Agropyron* vegetation of the Great Basin of western North America. It invaded the overgrazed semi-arid vegetation in the late 19th and early 20th century along transcontinental railroad tracks. By the 1930's it had largely replaced the perennial grasses under sagebrush and provided abundant dry tinder for great range fires which destroyed much of the remaining native vegetation. The situation was out of equilibrium in 1952 and it was stated "... the end is not yet in sight". It still is not in sight, and *Bromus tectorum* still is abundant and dangerous even though better fire control methods and planting of perennial range grasses have helped the situation.

The second example cited in 1952 concerned the accidental introduction from Eurasia of *Halogeton glomeratus* into the cold desert region of North America sometime before 1934. This opportunistic weed quickly spread along roadsides and into the drier overgrazed range lands of the northern intermountain region. *Halogeton* is poisonous to stock and millions of dollars have been spent trying to eradicate it. It seems to be well-adapted to its new environment and resists attempts to control it. Again, man has lost the battle with an invading species with a strong economic impact. *Halogeton* has continued to spread and now infests more than 10 million acres of dry rangeland (CRONIN, 1965). In view

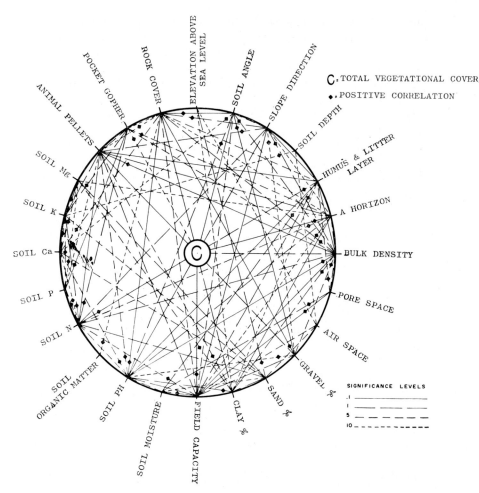

Figure 5. Statistically measured interactions between certain environmental components and between these components and vegetational cover in non-forested areas of north-western Yellowstone National Park. Not all possible interactions were tested. Diagram by courtesy of Dr. Duncan Patten of Arizona State University.

of the rapid evolution of heavy metal tolerance in plants (Antonovics, *et al.*, 1971), one wonders if the continued spread of both *Bromus tectorum* and *Halogeton glomeratus* into new environments in western North America may not be aided by natural selection of regional ecotypes from the gene pool of the species.

Newer examples of trigger factors are so abundant that only a few can be mentioned here. Certainly, among the most important and publicized are the various pollutants which have been added to the environment in ever-increasing amounts in the last twenty-five years or so. The pesticide DDT introduced in the 1940's has been shown to spread far beyond its original targets with its toxic residues accumulating in the higher trophic levels of ecosystems. Its use has now been severely restricted by the United States Environmental Protection Agency and, except for a few special uses where substitutes cannot be found, its use will be banned in the United States as of December 31, 1972. Its effects, however, on the continued survival of a number of species of carnivores in the earth ecosystem may never be overcome.

Since 1945, radionuclides of several kinds have become incorporated into the earth ecosystem. Again, these do damage at all trophic levels. Lower levels of nuclear testing and more research on the effects of these nuclides have and may continue to help in improving this situation. But much damage has already been done, and it must be said again that the end results of these additions to the ecosystem are not yet in sight.

Photochemical smog produced by the internal combustion engine now affects vegetation directly and indirectly in all but the subpolar and polar regions. Such smog is made up of so many compounds and the effects are so complex that solutions are not yet close. The Los Angeles smog now drifts hundreds of miles eastward from its source on the California highways. It has weakened and killed many trees of *Pinus ponderosa* in the southwestern mountains. The death of many of these smog-weakened trees is actually accomplished by a pine beetle to which healthy trees are fairly resistant. The trigger factor effect continues from automobile exhausts through the holocoenotic ecosystem to pines killed by smog and beetles — and, of course, to man himself.

The northern coastal plain of Alaska, particularly in the region around Point Barrow, is characterized by wet, patterned ground, and peaty soils underlain at 30 to 50 cm. by permafrost. In the last two or three decades, tracked vehicles have been used in various parts of this region to haul loads across the frozen tundra. Now, the effects of these tracked vehicles and their casual "roads" are all too apparent. Wherever the soil has been compressed or torn by even a few trips, destruction of the insulating soil has allowed an accelerating melting of the permafrost. Once the upper layers of permafrost in these soils are melted, relatively warm thaw waters in June flow down the vehicle tracks into streams and lakes. The result is drastic erosion from the streams back into the tundra along the two parallel ruts produced by the vehicles (Figure 6). In places, these erosion channels are now almost 2 meters deep and erosion gullies have proceeded as much as 30 or 40 meters back into the uneroded tundra soils and vegetation. While such "wilderness" areas are remote, some controls must be set up before these ecosystems are badly damaged by another example of a man-imposed trigger factor. Around Barrow, fortunately, controls and elevated roads are being established — but, again, it may be too late.

Some trigger factors act very quickly, others slowly and almost imperceptably over the centuries. As an example of relatively fast action, FOSTER (1973) describes the unexpected results of heavy rains during the normally "dry" season of December to April in the lowland forests of Panama. In January, 1970, these forests were subjected to very heavy rains for a few days, an unusual situation at that time of year. These rains did not greatly affect the forest plants (including trees and shrubs) which normally bloom at the end of the dry season and produce the first crops of seeds and fruits at the beginning of the wet season. However, the unusual rains did cause those plants which bloom and fruit in the wet season to bloom far too early or not at all. The result was a

Figure 6. Drastic and irreversible damage to tundra vegetation and soil due to melting of permafrost and consequent thermokarst erosion several years after the passage of a vehicle. Near Point Barrow, Alaska, at 71°N latitude.

lack of fruits and seeds in the forest late in the wet season (October, for example). This lack of fruits and seeds, of course, impeded the reproduction of many species. But it caused more serious trouble among many animal species; many animals died, probably of hunger. Others ate plants not normally eaten, such as the spiny palm. The buds were eaten out of many of these palms and thus the palms died. Other animals became adept at raiding the mess hall at the Smithsonian Laboratory on Barro Colorado Island and monkeys ran off with loaves of bread and other food. So, a few days of rain at the wrong time threw the whole ecosystem out of adjustment.

As examples of slowness of some trigger factors to act, ARON & SMITH (1971) cite the Erie, Welland, and Suez Canals. The Erie Canal was opened in 1819, the Welland in 1829, and the Suez in 1869. Migration of the alewife (*Alosa pseudoharengus*) into the lakes by way of the Erie Canal was not observed until 1873, and the sea lamprey (*Pentromyzon marinus*) not until the 1880's, and then only in Lake Ontario. It was not until 1921 that the lamprey was first caught in Lake Erie, getting there by way of the Welland Canal. Both of these invading species have had catastrophic effects on the populations of native and valuable fish species. The Suez Canal has resulted in almost a one-way migration of Indo-Pacific elements into the Mediterranean. The first observations of these elements was recorded in 1882 and by 1970 the list of these Indo-Pacific animal species in the Mediterranean had grown to 140, while only 42 species had moved in the opposite direction. The economic problems concerned with these migrations cannot be measured with accuracy. However, it can be said that equilibrium has not yet been achieved in the Great Lakes after 150 years and in the Mediterranean after more than a century since the opening of their respective canals triggered the migrations. Up to now, the Panama Canal has not caused such an impact because of the resistance provided by the locks and freshwater Gatun Lake. But maybe it is too soon to say.

The above examples are but a small fraction of the evidence accumulating on the ease with which whole ecosystems may be diverted or destroyed by changes in seemingly unimportant environmental factors. Referring again to the intricate relationships within the ecosystem, as oversimplified in Figure 2, we should be aware of the importance of knowing ecosystem pathways not only in small local situations but in the whole earth ecosystem itself. One has only to read the reports "Man's Impact on the Global Environment" (SCEP, 1970) and "Inadvertent Climate Modification" (SMIC, 1971) to realize that man's activities in the past have created countless problems for today whose solutions are not easy, if indeed they are possible of solving.

As steps in the solutions of environmental problems, two large projects, at least, must be undertaken. The first is a complete, world-wide monitoring of environments, biological communities, and ecosystems, including man. The second step is to develop predictive mathematical models of environmental component trends, whole environments, and whole ecosystems. The mathematical tools, computers, programmers, modellers, ecologists, and some data are at hand. Through various governments, agencies, and research programs, the first step is being taken. Some progress has been made on the more difficult second step but much remains to be done. As an example of a model of an environmental component which is changing due to man, we might look to that presented by MACHTA (1972), with the help of G. WOODWELL, D. KEELING, & J. OLSON, on the predicted and actual increase in atmospheric CO_2 from ca. 293 ppm in 1860 to 322 ppm in 1971, and the predicted 380 ppm in the year 2000. Observations at the Mauna Loa Observatory at 11,150 ft. elevation in Hawaii fit the model almost perfectly. The increase in CO_2 comes almost entirely from the burning of fossil fuels. If this release does not decrease quite soon, the CO_2 content will probably continue to fit the model. What will the biological and climatic consequences be? These, too, must be modelled and corrective steps taken.

Finally, through the International Biological Programme, whole ecosystems are being analyzed and predictive models are in the process of being constructed. Such computerized predictive models were not available to us in 1952. While these models are still relatively primitive, new data and revisions in the models are being made daily on the computers. It is too early to tell how soon these models will be useful but compared to our lack of knowledge and lack of tools twenty years ago, such efforts are far advanced and come none too soon.

REFERENCES

AL-ANI, H. A., STRAIN, B. R. & H. A. MOONEY, – 1972 – The physiological ecology of diverse populations of the desert shrub *Simmondsia chinensis*. *Journal of Ecology* 60: 41–57.

ALLEE, W. C. & T. PARK, – 1939 – Concerning ecological principles. *Science* 89: 166–169.

ANTONOVICS, J., – 1971 – The effects of a heterogeneous environment on the genetics of natural populations. *American Scientist* 59: 593–599.

ANTONOVICS, J., BRADSHAW, A. D. & R. G. TURNER, – 1971 – Heavy metal tolerance in plants. *Advances in Ecological Research* 7: 1–85.

Aron, W. I. & S. H. Smith, – 1971 – Ship canals and aquatic ecosystems. *Science* 174: 13–20.
Billings, W. D., – 1950 – Vegetation and plant growth as affected by chemically altered rocks in the western Great Basin. *Ecology* 31: 62–74.
Billings, W. D., – 1952 – The environmental complex in relation to plant growth and distribution. *Quarterly Review of Biology* 27: 251–265.
Billings, W. D., Godfrey, P. J., Chabot, B. F. & D. P. Bourque, – 1971 – Metabolic acclimation to temperature in arctic and alpine ecotypes of *Oxyria digyna*. *Arctic and Alpine Research* 3: 277–289.
Billings, W. D., Godfrey, P. J. & R. D. Hillier, – 1965 – Photoperiodic and temperature effects on growth, flowering, and dormancy of widely distributed populations of *Oxyria*. *Bulletin Ecological Society of America* 46: 89.
Billings, W. D. & H. A. Mooney, – 1968 – The ecology of arctic and alpine plants. *Biological Reviews* 43: 481–529.
Björkman, O., – 1968 – Further studies on differentiation of photosynthetic properties in sun and shade ecotypes of *Solidago virgaurea*. *Physiologia Plantarum* 21: 84–99.
Björkman, O. & P. Holmgren, – 1963 – Adaptability of the photosynthetic apparatus to light intensity in ecotypes from exposed and shaded habitats. *Physiologia Plantarum* 16: 889–914.
Bradshaw, A. D., – 1964 – Inter-relationship of genotype and phenotype in a varying environment. *Scottish Plant Breeding Station Record* 1964: 117–125.
Bradshaw, A. D., McNeilly, T. S. & R. P. G. Gregory, – 1965 – Industrialization, evolution and the development of heavy metal tolerance in plants. In "Ecology and the Industrial Society", *British Ecological Society Symposium* 5: 327–343.
Cain, S. A., – 1944 – Foundations of Plant Geography. Harper and Brothers, New York, 556 pp.
Chabot, B. F., Chabot, Jean F. & W. D. Billings, – 1972 – Ribulose-1,5-diphosphate carboxylase activity in arctic and alpine populations of *Oxyria digyna*. *Photosynthetica* 6(4): 364–369.
Christopherson, J., – 1967 – Adaptive temperature responses of microorganisms. In "Molecular Mechanisms of Temperature Adaptation", ed. C. L. Prosser, *American Association for the Advancement of Science Publication* No. 84: 327–348.
Clarkson, D. T., – 1965 – Calcium uptake by calcicole and calcifuge species in the genus *Agrostis*. *Journal of Ecology* 53: 427–435.
Cronin, E. H., – 1965 – Ecological and physiological factors influencing chemical control of *Halogeton glomeratus*. *Agricultural Research Service, U.S. Dept. of Agriculture Technical Bulletin* No. 1325. 65 pp.
Foster, R. B., – 1973 – Seasonality of Fruit Production and Seed Fall in a Tropical Forest Ecosystem in Panama. Ph.D. dissertation, Duke University, Durham, North Carolina, 156 pp.
Friederichs, K., – 1927 – Grundsätzliches über die Lebenseinheiten höherer Ordnung und den ökologischen Einheitsfaktor. *Die Naturwissenschaften* 15: 153–157, 182–186.
Friederichs, K., – 1957 – Der Gegenstand der Ökologie. *Studium Generale* 10. 112–144.
Friederichs, K., – 1958 – A definition of ecology and some thoughts about basic concepts. *Ecology* 39: 154–159.
Good, R., – 1931 – A theory of plant geography. *New Phytologist* 30: 149–171.
Good, R., – 1964 – The Geography of the Flowering Plants. 3rd ed. John Wiley and Sons, Inc. New York. 518 pp.

Heslop-Harrison, J., – 1964 – Forty years of genecology. *Advances in Ecological Research* 2: 159–247.
Kruckeberg, A. R., – 1967 – Ecotypic response to ultramafic soils by some plant species of northwestern United States. *Brittonia* 19: 133–151.
Kruckeberg, A. R., – 1969 – Soil diversity and the distribution of plants with examples from western North America. *Madroño* 20: 129–154.
Machta, L., – 1972 – Mauna Loa and global trends in air quality. *Bulletin American Meteorological Society* 53: 402–420.
Mason, H. L., – 1936 – The principles of geographic distribution as applied to floral analysis. *Madroño* 3: 181–190.
Mason, H. L., – 1946 – The edaphic factor in narrow endemism. II. The geographic occurrence of plants of highly restricted patterns of distribution. *Madroño* 8: 241–257.
Mason, H. L. & Jean H. Langenheim, – 1957 – Language analysis and the concept *Environment*. *Ecology* 38: 325–340.
McKell, C. M., Perrier, E. R. & G. L. Stebbins, – 1960 – Responses of two subspecies of orchardgrass (*Dactylis glomerata* subsp. *lusitanica* and *judaica*) to increasing soil moisture stress. *Ecology* 41: 772–778.
McMillan, C., – 1959 – The role of ecotypic variation in the distribution of the Central Grassland of North America. *Ecological Monographs* 29: 285–308.
Merril, C. R., Geier, M. R. & J. C. Petricianni, – 1971 – Bacterial virus gene expression in human cells. *Nature* 233: 398–400.
Mooney, H. A. & W. D. Billings, – 1961 – Comparative physiological ecology of arctic and alpine populations of *Oxyria digyna*. *Ecological Monographs* 31: 1–29.
Mooney, H. A. & A. T. Harrison, – 1970 – The influence of conditioning temperature on subsequent temperature-related photosynthetic capacity in higher plants. In "Prediction and Measurements of Photosynthetic Productivity", ed. C. T. deWit, Centre for Agricultural Publishing and Documentation, Wageningen, The Netherlands: 411–417.
Mooney, H. A. & Marda West, – 1964 – Photosynthetic acclimation of plants of diverse origin. *American Journal of Botany* 51: 825–827.
Muir, J., – 1911 – My First Summer in the Sierra. Houghton Mifflin Company, Boston.
Odum, E. P., – 1963 – Ecology. Holt, Rinehart and Winston, New York. 152 pp.
Odum, H. T., – 1971 – Environment, Power, and Society. Wiley-Interscience, New York. 331 pp.
Olmsted, C. E., – 1944 – Growth and development in range grasses. IV. Photo-periodic responses in twelve geographic strains of side-oats grama. *Botanical Gazette* 106: 46–74.
Patten, D. T., – 1972 – Environmental relations of non-forest vegetation in northwest Yellowstone National Park. Unpublished manuscript, Arizona State University, Tempe, Arizona.
Perry, T. O., – 1962 – Racial variation in the day and night temperature requirements of red maple and loblolly pine. *Forest Science* 8: 336–344.
Prosser, C. L. (ed.), – 1958 – Physiological Adaptation. American Physiological Society, Washington, D.C. 185 pp.
Rune, O., – 1953 – Plant life on serpentines and related rocks in the north of Sweden. *Acta Phytogeographica Suecica* 31: 1–139.
Scott, D. & W. D. Billings, – 1964 – Effects of environmental factors on standing crop and productivity of an alpine tundra. *Ecological Monographs* 34: 243–270.
Semikhatova, Olga A., – 1970 – The after-effect of temperature on photosynthesis. *Botanicheskii Zhurnal* 45: 1488–1501.
Strain, B. R. & Valerie C. Chase, – 1966 – Effect of past and prevailing

temperatures on the carbon dioxide exchange capacities of some woody desert perennials. *Ecology* 47: 1043–1045.

TANSLEY, A. G., – 1917 – On competition between *Galium saxatile* L. and *Galium sylvestre* Poll. on different types of soil. *Journal of Ecology* 5: 173–179.

TEERI, J. A., – 1972 – Microenvironmental Adaptations of Local Populations of *Saxifraga oppositifolia* in the High Arctic. Ph.D. dissertation, Duke University, Durham, North Carolina. 216 pp.

TOBIESSEN, P. L., – 1970 – Temperature and Drought Stress Adaptations of Desert and Coastal Populations of *Isomeris arborea*. Ph.D. dissertation, Duke University, Durham, North Carolina. 142 pp.

TURESSON, G., – 1922 – The genotypic response of the plant species to the habitat. *Hereditas* 3: 211–350.

VAARTAJA, O., – 1959 – Evidence of photoperiodic ecotypes in trees. *Ecological Monographs* 29: 91–111.

WATSON, J. D. & F. H. C. CRICK, – 1953 – Molecular structure of nucleic acids: A structure for deoxyribose nucleic acid. *Nature* 171: 964.

WILSON, C. L. & W. H. MATTHEWS (eds.), – 1970 – Man's Impact on the Global Environment. *Report of the Study of Critical Environmental Problems (SCEP)*, MIT Press; Cambridge, Massachusetts. 319 pp.

WILSON, C. L. & W. H. MATTHEWS (eds.), – 1971 – Inadvertent Climate Modification. *Report of the Study of Man's Impact on Climate (SMIC)*. MIT Press, Cambridge, Massachusetts. 308 pp.

3 THE ECOLOGICAL NICHE AND VEGETATION DYNAMICS

James E. Wuenscher

	Contents	page
3.1	Introduction	39
3.2	Development of the Niche Concept	39
3.3	The Niche as a Vector Space	41
3.4.1	Summary	44

3   **THE ECOLOGICAL NICHE AND VEGETATION DYNAMICS**

3.1   **Introduction**

The term "ecological niche" has been used with increasing frequency in the ecological literature since its introduction by GRINNELL in 1917. The niche concept has been discussed from many points of view, and its meaning has continuously evolved. In nearly all cases, however, the niche concept has remained in the domain of theoretical speculation rather than serving any practical use.

This has been particularly true in the field of vegetation analysis, where the ideas of niche difference and competitive exclusion have been much talked about in explaining observed plant distributions and community composition but little used in elucidating the reasons for those phenomena. This paper, while still largely theoretical, attempts to clarify and expand the current niche concept and to suggest how it may actually be used in the analysis of vegetation and ecosystems.

3.2   **Development of the Niche Concept**

The concept of the ecological niche has been developed largely in studying the ecology of animals. Zoological use of "niche" has most often been based on the nutritional role of an animal in its ecosystem (WEATHERLEY, 1963). This usage had developed from ELTON's (1927) statement that "... the niche of an animal means ... its relation to food and enemies". While this usage has been of great value in zoological trophic dynamics, it is difficult to apply to plants. Plants are limited in their relations to food and enemies to being either primary producers, food sources, or decomposers, neither classification being specific enough to serve as definitions of their niches.

GRINNELL (1928), however, defined the "ecologic or environmental niche" more broadly as "the ultimate distributional unit within which each species is held by its structural and instinctive limitations". Thus, a species' niche is a functional unit within the biotic community (UDVANDY, 1959), including relations of the species to all of the resources of an ecosystem that it is capable of utilizing (DICE, 1952). The result of reasoning along this line was HUTCHINSON's (1958) now classic description of the niche as "... the sum of all the environmental factors acting on an organism; the niche thus defined is a region of n-dimensional hyperspace". This approach defines niche in terms of all variables relative to the species, regardless of their exact nature, and makes the niche concept directly applicable to plants as well as animals. Indeed, it was not until the advent of the hyperspace concept that the term niche became truly applicable to plants.

According to the niche hyperspace concept, the environment of an organism is a set of continuous variables, each represented as an axis of continuous variation between extremes. Each environmental axis is considered at some angle to every other environmental axis in a many-dimensional space. Along each axis any particular species has a tolerance range in which it can live successfully. By defining the tolerance range of a species along n environmental axes, a volume is generated in n dimensions that represents the potential survival area of the species. The niche is, then, an abstract environmental volume within which a species' responses are inside the critical boundaries of survival. A species' niche can be defined by quantifying species' responses to environmental variables and delineating the region of response within which the species is above or below a critical extreme. This region is essentially identical to GRINNELL's ultimate distributional unit.

This definition of the niche, however, leaves us with only the outlines of a species "potential" niche, that is, the region in environmental hyperspace in which the species could survive if its survival were limited only by its physiological tolerance. Clearly, some regions within the n-dimensional niche volume will be more favorable for the species than other regions. It is necessary to know these regions of high suitability if we are to understand the processes involved in determining the species "realized" niche, that is, the region of environmental hyperspace that a species actually occupies in nature when it is limited by competition. To fully define a species' niche it is necessary, then, to include not only the limits of survival of the species along environmental gradients, but the level of response within these limits. The niche hypervolume is then bounded not only by the limits of survival of a species, but also by its response to the impinging environment at all points along the environmental axes.

Complete specification of a species' niche would include every possible response to every environmental variable. Realistically this is impossible, and the best that can be achieved is an open-ended "partial" niche description (MAGUIRE, 1967). This practical limitation is not serious, however, since under natural conditions an organism's survival is determined by a few critical factors which actually impinge upon it and significantly influence its biological processes. These critical factors no doubt vary with the situation and species involved, but they should be identifiable with a relatively small amount of observation and experimentation. Once the number of significant environmental factors is limited, organism response to varying levels of these factors can be determined experimentally.

As an example, consider tree seedlings of several species growing on a range of sites from full sun exposure to deep shade. A species successful at one extreme of this gradient presumably has a different response volume, or niche space, from a species successful at the other extreme. Photosynthetic and transpirational responses to varying levels of incident radiation and moisture conditions likely to occur along the environmental gradient can be measured. If a seedling can achieve a near optimal level of carbohydrate production and maintain adequate cell water potential, its chances of survival are good. Tolerance ranges can then be defined as the range in which a species maintains a positive photosynthesis-respiration balance and does not become dehydrated. The volume defined by the axes of environmental variation and species responses will partially specify the species niche. If the environmental gradients selected are, in fact, those critical for differential survival of the species involved, the species' partially defined niches will consist of different volumes within the environment-response hyperspace. This follows WHITTAKER's statement (1967) that "The community is an assemblage of niche-differentiated species which conceivably, if ecologists understood enough, might be ordinated in a niche hyperspace".

3.3 The Niche as a Vector Space

Even if limited to a few critical dimensions, the environment-response hypervolume comprising a species niche is difficult to work with. The human mind has difficulty conceptualizing beyond the three dimensions it knows by direct experience. We can, however, use a technique for niche specification that has been widely used in the physical sciences to allow analysis in more than three dimensions. By considering the niche hyperspace as a many-dimensional vector space, it is possible to specify the niche quantitatively in a readily-usable way.

Any point within the niche is a function of all environmental variables and all response variables. This point is then a vector within the vector space and can be expressed in vector notation as a one-dimensional matrix, $P_1 = (E_1, E_2, E_3, \ldots R_1, R_2, R_3, \ldots)$, where E_1 is a specific value of some environmental variable, and R_1 is the organism response corresponding to that environmental input. The niche is the set of all environmental variables (essentially the habitat) and all organism responses, and both the habitat and total response are subsets of the niche. The habitat corresponding to a specific point in the niche is represented by the vector $(E_1, E_2, E_3, \ldots)$, and the organism response by the vector $(R_1, R_2, R_3, \ldots)$. This helps clarify the distinction between habitat and niche, two terms that have often been used to mean much the same thing.

We can consider the entire field of environmental and response variation as the total niche-space, within which each species' niche occupies some particular volume. By examining the relationships among species' niches within niche-space, we can define niche overlap and differentiate between potential and competitively realized niches. If there are no habitat vectors in common, the species' niches do not overlap, and they do not compete. If the habitat vector subsets of species' niches overlap, competition will occur between them in the region of overlap. The successful competitor will be determined by the species' response vectors.

Response vectors differ in different parts of a species' niche as biological processes respond to environmental gradients. There exists an optimal part of a species niche with suboptimal conditions in other areas. The most successful competitor for a particular environmental vector will be the species whose response vector is most nearly optimal at that point. The possibility of extending this analysis to predictive modeling of competition has been discussed previously (WUENSCHER, 1969).

The discussion up to this point has implied a static condition, that is, a species' location within niche space has been assumed to remain constant. But one of the major characteristics of living systems is that they are dynamic and change through time. Consideration of the niche in analysis of vegetation dynamics requires that its behavior through time be examined. At least three time scales are relevant: (1) that of the life cycle of an individual organism as its niche volume changes through time, (2) that of a community or ecosystem as it matures, and (3) that of evolutionary time as adaptive changes bring about shifts in the location of the niche hypervolume.

Throughout the life cycle of a given organism, the location

of its niche hypervolume within niche-space will change. Some of the variations may be directional and non-reversible such as changes due to maturation and aging. Some, on the other hand, will be cyclical due to diurnal or seasonal changes and the effects of environmental preconditioning. All of these changes may be visualized as pulsations and drifting of the niche volume through niche space. The degree and direction of drift will depend upon the extent to which the response and habitat subsets of the species niche change through its life-cycle.

As a plant community matures, it too can be thought of as expanding and contracting and drifting through niche space. That part of niche space occupied by a plant community consists of the total volume of the niches of all the species within the community. Since the niche habitat subsets of the species within a single community will be similar, their niche hypervolumes would be expected to be mostly contiguous. The entire community, then, would occupy some specific area of niche space without marked discontinuities. This area can be considered as the niche hypervolume of the community.

Succession within the plant community can be represented as movement through niche space. As the environment of the community changes it moves along the axes of environmental variation and eventually passes beyond the bounds of optimal survival of the species present. Then new species move in whose responses are more nearly optimal in the part of niche space the community now occupies. The resulting movement of the plant community through the abstract field of niche-space will presumably be, in the absence of drastic disturbance, continuous and directional, or at least non-random. Thus succession is subject to mathematical description in the vector field of niche space.

An interesting side-point that may be considered here is whether the direction of change in the plant community is toward stability or whether it is continuous or cyclical, never reaching a stable point. Rather than defining stability as constancy of environment or of species composition, we can define it as maintenance of the community within a certain area of niche space. It is possible that niche-space stability may be maintained without necessarily maintaining a constant species composition, or vice versa.

Changes in diversity as a community matures may also be considered within the context of niche-space. It is often assumed that as a community matures it increases in species diversity and complexity. Increase in species diversity is dependent upon the occurrence of one or both of two possibilities: (1) an increase in habitat diversity or (2) a decrease in niche size. Increasing habitat diversity would necessitate an increase in community size in niche-space with the passage of time. A decrease in niche size of the individual species, however, would mean that the community could occupy the same or even a reduced volume in niche space and still contain more species. Analysis of community succession in niche-space could shed light on the extent to which these processes occur.

The niche hyperspace also provides a framework for the analysis of evolutionary processes. The evolution of species' niches has received considerable attention from population biologists and geneticists (LEVINS, 1968). In vegetation dynamics, the relationships between species niches as they evolve in niche-space is perhaps more pertinent than the evolution of the individual species, themselves. The controversy over the continuity or discontinuity of plant communities could be largely resolved if we knew whether species have evolved away from one another in niche-space as proposed by WHITTAKER (1967) or if they have evolved toward one another to form distinct clusters. If the former possibility were true, we would expect the major portion of the niche-hyperspace to be filled by different species niches. This situation would best be described by the continuum concept. If species have evolved together to form community clusters, however, sharp discontinuities would exist with niche-space and also in the vegetation of a region. Both situations appear to exist in nature.

3.4.1 SUMMARY

The expansion and modification of the Hutchinsonian niche concept described in this paper provides a useful framework for approaching several aspects of plant ecology. Initially it provides a means of functionally relating organism response to environmental variables. Thus, organism behavior at a given time under specific conditions can be specified in discrete numbers instead of vague generalizations, making the niche concept usable in a real sense. The measurements made by "physiological" ecologists can be applied to vegetation dynamics through niche specification and analysis.

In addition, use of the concepts and techniques of vector space analysis provides a method for synthesis of population genetics, physiological ecology, and vegetation dynamics. The n-dimensional vector field approach is one of the most fundamental mathematical techniques available for consideration of transformations of complicated systems in time. Thus it is becoming more and more widely used by biologists in many fields (LEWONTIN, 1969; WATT, 1968). The modern niche concept thus provides a theoretical and mathematical approach to ecology that should enable us to better understand and hopefully to predict vegetation and ecosystem dynamics.

REFERENCES

DICE, L. R. – 1952 – Natural Communities. Univ. Mich. Press. Ann Arbor.
ELTON, C. S. – 1927 – Animal Ecology. Sidgwick and Jackson. London.
GRINNELL, J. – 1917 – The niche-relationships of the California thrasher *Auk* 34: 427–433.
GRINNELL, J. – 1928 – The presence and absence of animals. *Univ. Calif. Chron.* 30: 429–450.
HUTCHINSON, G. E. – 1958 – Concluding remarks. *Cold Spring Harbor Symp. Quart. Biol.* 22: 415–427.
LEVINS, R. – 1968 – Evolution in Changing Environments. Princeton Univ. Press. Princeton, N. J.
LEWONTIN, R. C. – 1969 – The meaning of stability. *Brookhaven Symposia in Biology* 22: 13–24.
MAGUIRE, B., JR. – 1967 – A partial analysis of the niche. *Amer. Nat.* 101: 515–526.
UDVARDY, M. – 1959 – Notes on the ecological concepts of habitat, diotope, and niche. *Ecology* 40: 725–728.
WATT, K. E. F. – 1968 – Ecology and Resource Management. McGraw-Hill. N.Y.
WEATHERLEY, A. H. – 1963 – Notions of niche and competition among animals, with special reference to freshwater fish. *Nature* 197: 14–17.
WHITTAKER, R. H. – 1967 – Gradient analysis of vegetation. *Biol. Rev.* 42: 207–264.
WUENSCHER, J. E. – 1969 – Niche specification and competition modeling. *J. Theoret. Biol.* 25: 436–443.

4 DESCRIPTION OF RELATIONSHIPS BETWEEN PLANTS AND ENVIRONMENT

D. Scott

Contents

4.1	Introduction	49
4.2	The Plant	49
4.3	Environmental Variables	50
4.4	Relationships between Variables: Causal Diagrams	54
4.5	Ordination and Classification of Data	56
4.6	Fitting of Functions	57
4.7	Statistics and Linear Relationships	59
4.8	Curvilinear Relationships	64
4.9	Conclusions	68

4 DESCRIPTION OF RELATIONSHIPS BETWEEN PLANTS AND ENVIRONMENT

4.1 Introduction

The descriptive and functional characteristics of a vegetation result from interactions between the properties of the plant species it contains and the environment in which they occur. The problem is to find which processes and interactions are important in a particular situation and to estimate the magnitude of the effects. The following paper gives some concepts and methods of data analysis which may be used.

At this stage, the reader is reminded that such a paper is as likely to reflect the personal opinions and biases of the author, as it is to be a general review of the subject; and, in common with many aspects of science, it is likely to reveal more problems than solutions.

4.2 The Plant

The existence of a living plant demonstrates two things about its environmental relationships. Firstly, that none of the environmental variables, either alone or in combination, have exceeded the tolerance range of the plant at any stage during its lifetime; and secondly, that while growth has presumably varied with environmental fluctuations, the plant has made net growth. Thus, there are two types of analytical problems; the first concerned with defining the tolerances, and the second, the performance (*e.g.*, growth) under the fluctuating environmental conditions within these tolerances. The potential tolerances and responses are determined by the genome of the plant, but its response at a particular instant will be conditioned by its past environmental interactions (Fig. 1).

Because of the known difference in response between plants, the analysis of a vegetation/environment complex should probably be attempted at the level of the species or appropriate physiological-genetic entity. With this approach, the total vegetation/environment relationships would be a synthesis of the species/environmental

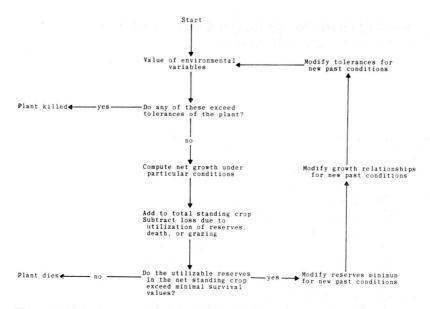

Figure 1. Flow diagram illustrating response of a plant to fluctuating environmental conditions in successive time intervals (based on Scott & Billings, 1964).

relationships, and will tend to emphasize the interaction between the species within a vegetation. The alternative holistic approach to vegetation/environment relationships has been discussed by Billings earlier in this chapter.

Environmental variables differ in the type and magnitude of their effect on plants. But the response of the plants will be conditioned by 'all' environmental variables. Thus, in studying the relationship between plants and particular environmental variables one must take account of possible influence of other variables. For example, in the study of the relationship between a species and soil properties one should find out whether or not the species is excluded from a portion of the sequence by competitive exclusion caused by a second species.

4.3 **Environmental Variables**

Environmental variables can be defined as any aspect or process surrounding the plant or vegetation, the levels or fluctuations of which, influence the plant at some stage during its lifetime (Major, 1951; Mason & Langenheim, 1957; Billings (this part)).

TABLE 1

Aspects of environmental variables that have been considered in vegetation studies and their principal effects on plants (modified from BILLINGS, 1952). A, Radiation regime; B, Temperature or energy exchange regime; C, Nutrient regime; D, Water regime; E, Living regime in time or space or factors controlling plant responses

Variables	Aspects	Plant regime
	Climatic	
Radiation solar,	Spectral composition, intensity, direction, duration, periodicity	A, B, D, E
terrestrial	Intensity, integration	B
Temperature, air	Degree, periodicity, integration, lateral & vertical variation	B, D, E
soil	Degree, periodicity, lateral & vertical variation, freeze – thaw phenomena	B, D, C, E
Water, vapour	Amount or densities, vapour pressures & VP deficits, evaporation, transpiration	D, B
condensed water	Cloud, fog, dew, salt spray	D, C
precipitation	Types, amounts, frequencies. droughte. snow cover, snow abrasion	D, E, C, E
soil water	Content, tensions, supply rates, soil aeration	D, B, C, E
Atmospheric gases composition	Oxygen & carbon dioxide contents and partial pressures, ozone, pollutant gases	C, E
pressure	Weather phenomena, gas partial pressures	A, B, D, C
wind	Frequency, force, direction, evaporation, abrasive agents	E, D
	Edaphic	
Parent material	Minerals present, structure, weathering susceptibility	C, D
Soil, physical properties	Profile, structure, texture, mechanical analysis, soil moisture, stability	C, D, B
chemical properties	Clay mineralogy, organic compounds & breakdown, base exchange properties, pH, macro- & micro- nutrients, toxicity elements	C
biotic properties	Soil flora & fauna, litter & humus, breakdown rates, antibiotic effects	C, E
	Geographic	
Position latitude, longitude, etc.	Work through other factors e.g. climate, erosion	A, B, C, D
Topography slope, aspect, altitude	Work through other factors e.g. climate, erosion	A, B, C, D
Vulcanism thermal effects		B
mechanical	Ash cover, lava flows	E
	Pyric	
Climatic	Temperature (air & soil), intensity & duration, post burning microclimate effects	B

Edaphic	Organic matter destruction, mineral release, erosion	C, D, E
Biotic	Community composition	E
	Biotic	
Other plants	For light, water & nutrients, antibiotic &	
competition	autotoxic effects	A, B, C, D, E
dependence	Litter & humus, physical & chemical effects, cover	A, B, C, D, E
Animals		
destructive	Grazing, feeding, etc., effect on soil	E
beneficial	Nutrient cycling, fruit & seed dispersal	C, E
Man	Can change almost any factor, at least locally	A, B, C, D, E

TABLE 2

Relationship between statistical methods in terms of nature and distribution of measured variables.

Method	Independent						Dependent			
	unmeasurable	qualitative	qualitative & quantitative	quantitative	independent & normal dist.	independent & random	may be correlated	qualitative	quantitative	normal dist.	not necessarily
Path Analysis				*			*		*		*
Multiple Regression (Type I or fixed model)				*	*				*		*
Multiple Regression (Type II or random model)				*	*				*	*	
r (Simple Correlation)				*	*			ND	ND		
Multiple Regression (Multiple Correlation)				*	*			ND	ND		
Partial Correlation				*	*				*	*	
Discriminant Analysis				*	*			*		*	
Covariance Analysis			*	*				*		*	
Analysis of Variance (Type I or fixed model)		*			*			*		*	
Analysis of Variance (Type II or random model)		*			*			*		*	
Contingency Table		*			*			ND	ND		
Principal Component Analysis	I				*			*		*	
Factor Analysis	I					*		*		*	
Canonical Correlation	I					*		*		*	

ND = No Distinction between dependent and independent variables
I = Inferred during analysis

Environmental variables are often listed under the headings of climatic, edaphic, biotic, pyric, and topographic. But in relation to their effect on plants, a functional subdivision could be in terms of the radiation, — temperature, — moisture, — nutrient, — and biological regimes of the plant. The types of environmental factors that have been measured in vegetation studies have been listed by BILLINGS (1952, 1974) and in Table 1.

The values of each of the environmental variables may change over time scales or periodicities of minutes, hours, days, years and centuries, with their associated effects on plants (SALISBURY et al., 1968).

The environmental variables that are relevant to a particular problem are determined by several considerations. First there are the objectives of the study. For example, consider the effect of light on vegetation. If the problem were the broad geographic distribution of vegetation on a global scale, then mean annual radiation figures might be sufficient. Whereas, if the problem were on the day-to-day growth and differentiation of a plant, it would be necessary to consider the influence of radiation intensities, spectral composition, and periodicities. Thus, different problems will require working at different levels of complexity, and what is in one investigation a simple environmental variable, may in another require subdivision into components.

Second, there is the point of view of the investigator. A pedologist may view vegetation and animals as causative factors with respect to soil properties, whereas, to a plant ecologist, soil and animals, in the same situation, may be regarded as causative agents determining vegetational properties. Thirdly, the same example illustrates the time scale aspect. Over a period of hours or days it may be reasonable to consider soil properties as relatively stable causative agents with respect to vegetation behaviour, whereas over years or centuries, vegetation is one of the main factors influencing soil properties.

Fourthly, there may be a difference in emphasis depending on the objectives of the study. If one were concerned with predicting timber yield from a number of environmental variables then the main criteria in one instance could be a predictive ability of the relationship irrespective of the reasons for the relationship. Alternatively, understanding the basis of the relationships may be the objective — though this will generally require measurements and understanding of more variables.

The number of environmental variables requiring measurement in a particular problem should be overestimated rather than underestimated. This is particularly so where mathematical methods are used because while mathematical methods may be able to exclude variables because of their lack of quantitative effect, they cannot include unmeasured but important variables. While precise measurements are preferred, even subjective scalars based on the field ecologists experience and assessment are better than no measurements.

Thus, while an approach considering all plant and environmental variables, all their interactions, over all time scales, may be a desirable objective, in practice it will usually be replaced by approximations suitable for the purpose and time span of the particular problem.

4.4 Relationship between Variables: Causal Diagrams

The number of plant and environmental variables present in a particular problem often pose conceptual and analytical problems. The method of causal or structural diagrams can be useful in both regards.

In most problems some variables can be recognized as causes and others as effects, and a variable which is an effect in one relationship may be a cause in another.

This can be shown diagrammatically where the variables are listed and where arrows indicate the relationship between variables. For example the relationship between variables A, B, C and D might be:

where C is influenced by variables A and B, while variable D is influenced directly by variable B and C and only indirectly by variable A. In such a diagram, a variable is interpreted as being a function of only those variables at the tails of those arrows leading to it. The arrow indicates the pathways of influence.

The points which can be shown by such a diagram are, firstly, that some variables (A & B) are primary causes in that they are uninfluenced by other variables within the scheme. Secondly, that other variables (C) can be intermediate effects, in being the effect in relation to one set of variables (A & B), but in turn are causes in relation to further variables (D). Thirdly, a variable may exert its influence via different pathways, e.g., B has a direct influence on D as well as in indirect influence through its effect on variable C.

There is no limitation to the number of variables and number of interactions that can be considered in this multi-stage manner (Fig. 2).

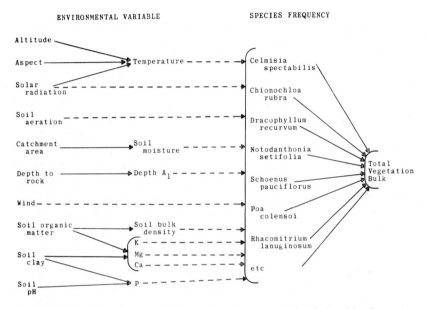

Figure 2. Example of causal diagram showing expected relationships between measured vegetation and environmental variables in a particular study in a New Zealand alpine area.

Such a scheme will be drawn up from the investigator's knowledge of the expected relationships between the variables — either known or hypothesized. In this, he well need to utilize information from all available sources. The arrows in such diagrams may have interpretations other than cause and effect; for example they may also imply the passage of time.

BILLINGS (this part) discussed the holistic approach with its indication of the many potential variables and interactions. However, these interactions arise because of biological or physical processes linking particular variables, *e.g.*, wind by increasing the turbulent transfer of heat and water vapour away from a leaf. Any particular variable will only be directly effected by a few variables, and in turn will only have a direct effect on a few further variables. More complex indirect interactions are combinations of these simpler direct relationships. The causal diagram approach, while not making a problem any less complex, makes it manageable in that attention can be directed as each link in the scheme as required.

4.5 **Ordination and Classification of Data**

Where there are sets of measurements of plant and environmental variables, any relationship between the two should be apparent as patterns within the data. However, at least in the early stages of many problems, these patterns may not be obvious. There are several techniques concerned with showing the relationship or groupings amongst sets of data, *e.g.*, the graphical ordination methods of BRAY & CURTIS (1957), similarity coefficients of JACCARD (1932), and more recently the agglomerative and devisive methods using qualitative and quantitative variables in vegetation and taxonomic problems (*e.g.*, LANCE & WILLIAMS, 1965 & 1967; RANDALL, 1969). These generally have been applied to vegetation data, less commonly to environmental data (*e.g.*, TRACY, 1969), but seldom to both.

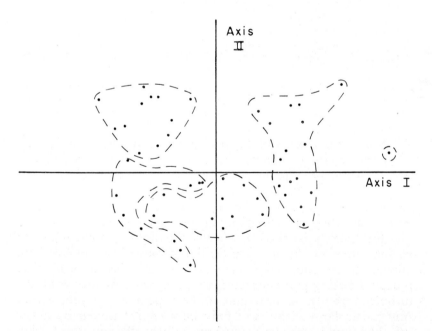

Figure 3. Ordination of some Australian forest types based on floristic analysis, with the points encircled to indicate the corresponding classification of sites based on soil physical properties (TRACY, 1969).

In vegetation/environment studies, these methods are usually used to classify or group the sites or stands on the basis of vegetation data and then from the associated environmental data to find which factors are characteristic of each grouping (Fig. 3). (See also the paper by JON SCOTT in this chapter.) While such methods may indicate the important environmental variables they do not in themselves give the quantitative expressions for individual relationships between environmental variables and plants.

4.6 **Fitting of Functions**

The information on the relationship between two or more variables is likely to progress through several stages (Fig. 4). Initially, it may be possible to make only qualitative statements (Fig. 4A). Some relationships will only be of this type. For example, if days are longer than a critical length then certain plants come into flower. The second stage (Fig. 4B) can be used where there are quantitative measurements of the variables and where the relationships are presented in tabular or graphical form for the individual points, or as free-hand graphs. These are the most common form of data and data presentation. The remaining two stages use mathematical equations fitted to sets of quantitative data by statistical or other mathematical means. A distinction is made

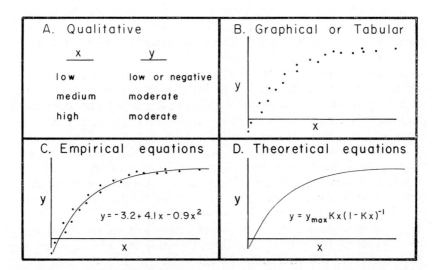

Figure 4. Stages in the understanding of relationship between two variables – example is the relationship between net photosynthesis (y) and light intensity (x)

between empirical and theoretical equations (Fig. 4C and 4D). Most of the discussion will refer to these latter two.

Mathematical equations provide a shorthand way of describing the relationship between variables. Empirical equations are those in which the mathematical form of the equation is determined solely by its ability to reproduce the relationship between the measurements shown in the sample data; the parameters in the equations may have no importance apart from this. While such empirically fitted equations simplify the utilization of the data there is a danger that the unwary, impressed with their mathematical "elegance", will place undue importance to them. Also there is no justification for extrapolating such relationships beyond the range of values for which they were established.

Theoretical equations are those which have been derived from consideration of the mechanisms involved and in which each parameter has a definite interpretation. Such equations will generally include coefficients which will have to be determined from the quantitative data of the particular problem. Such equations are possibly capable of extrapolation.

The chief criterion of the applicability of particular types of equations are the extent to which they simulate the relationship between the variables. The general form of these is given (Fig. 5).

In the first (Fig. 5A), the plant or other characteristic is present only within a certain range of the environmental factor and absent from the rest of the range. Within the range in which it is present, the characteristic may vary from zero at the lower threshold, rise to a maximum at some intermediate level, and depress to zero again at an upper threshold. The exact shape of the relationship depends on the particular situation. The environmental variable may continue to vary outside the range in which the plant characteristic is present. In relation to fitted mathematical equations, it is important to note that the value of the plant characteristic in that environmental range is zero and not a negative or complex number.

In the second (Fig. 5B), the plant or vegetation characteristics (*e.g.*, growth rate) have both upper and lower limits as well as upper and lower thresholds. The types of relationship within these limits may be diverse with direct or inverse relationships or with maxima or minima occurring at some intermediate levels. The first is only a specific case of the second more general relationships.

Where there are many variables, the same type of relationships will occur in the multidimensional analogy. In all cases, the objectives will be define the limits, and the exact form of the relationship within these limits.

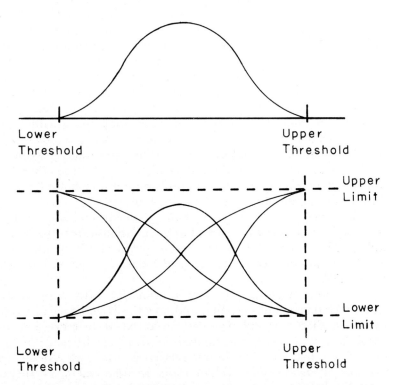

Figure 5. General form of relationship between vegetation or plant characteristics and one environmental variable. Upper curve is Fig. 5A; lower curves are Fig. 5B.

Suffice to say there is no general class of mathematical equations which is sufficiently flexible to accommodate all these possibilities. Most investigations are concerned with only a segment of the full response curve.

4.7 **Statistics and Linear Relationships**

The fitting of a function or equation involves two stages — first the selection and fitting of an appropriate equation, — and second the determination of the reliability or significance of these equations or their parameters.

The previous discussions presupposed that the particular relationship between two or more variables was known to be present and important. The more usual situation is for the relationship to be anticipated or hypothesized and the problem to be one of determining whether the effect exists, and its magnitude, in a particular set of data, after the effect of other variables, chance effects, or sampling errors had been taken into account. Statistical or probabilistic methods are required where such decisions have to be made. Accordingly statistical methods are likely to be particularly important at the qualitative or empirical quantitative stages.

There are many statistical methods in use but only some of the common ones will be discussed. The basic requirement of any statistical method will be some element of randomness with regard to the selection of the sample data. But the requirements differ between methods. A consideration of the nature and distribution of the values of the variables show that each method has been developed for a particular type of problem but in combination they form a continuum of methods (SCOTT, 1969).

The methods are listed in Table 2 in approximate order of their ability to provide quantitative relationships between variables. The other subdivisions are in terms of the types of data available and the requirements of the data if statistical tests are to be made.

In most methods, interest is primarily with one variable (subject or dependent variable) and how this is influenced by other variables (factor or independent variables) as in regression, discriminant, covariance, and analysis of variance. In path analysis, the independent variables in one relationship may be a dependent variable in an associated relationship. In other methods, there is no distinction between dependent and independent variables and attention is directed at the relationship between the group as a whole (simple and multiple regression, contingency tables, principal component, factor analysis and canonical correlation).

All the methods listed, with the exception of contingency tables, assume a linear additive relationship between the variables. For example, where there is a single dependent variable and several independent variables of the form

$$y = b_0 + b_1 x_1 + b_2 x_2 + \text{etc.} + (\text{error or residual effect})$$

where the dependent variable (y) is the measured plant of vegetation characteristic, the x's the measured environmental variables, and the b's fitted coefficients. Graphically these are straight lines or planes. The values of the fitted coefficients are such that there is a least squares deviation between the actual and predicted values of the dependent variable. However, while these linear equations are implicit in the methods listed they are generally only explicitly given in path —, regression —, and discriminant analysis. In the last three methods listed, the equations are given for the derived variables in terms of the measured variables.

Another subdivision is in terms of the type of data. Several methods use data in which all variables have been measured as continuous quantitative variables (all in first group). In other methods some of the variables may be qualitative in either the dependent variable (discriminant analysis); in some (covariance analysis) or all (analysis of variance) of the independent variables; or in all variables where there is no distinction (contingency tables). This distinction between the measured variables being quantitative or qualitative in somewhat artificial, for in the computation, the qualitative variables are treated as quantitative variable by coding (e.g. 1 = presence of qualitative character, 0 = absence of character). The independent variables in the last three methods are listed as unmeasureable in the sense that they are derived by the relationships between the measured variables.

Any statistical analysis will depend on knowledge of the distribution of values in a random sample of the variables. The Normal or Gaussian distribution figures prominently in the methods listed. But the point at which randomness and normal distribution apply differ between the methods.

The minimum requirement in most methods is that the error or residual effects of the fitted relationship be random and normally distributed. In addition, some assume that the measured values of the variables are also random samples from normally distributed range. This applies to all variables in simple —, partial —, multiple correlation, discriminant —, covariance analysis, and Type I or random multiple regression and analysis of variance (multivariate normal distributions). In other methods, the values of the independent variables may have been selected, though there is a difference depending on whether the independent variables as their name suggested vary in value independent of each other, whereas in others (path analysis, factor analysis) they may be correlated. The variables for contingency tables are listed as independent in the sense that this is the null hypothesis that one wishes to disprove.

In all cases, these random or normality characteristics of the variables or their departures should be investigated and known

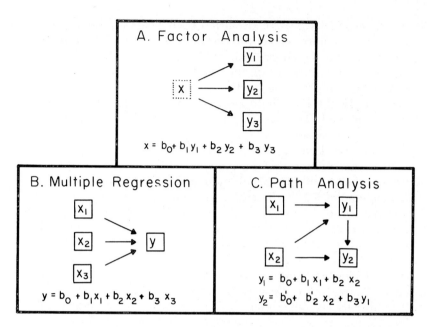

Figure 6. Causal diagram representation of three statistical methods and associated fitted equations.

and not just assumed as is commonly done.

Three of the methods will be discussed in terms of the causal diagram approach and the empirical equation which the method fits to the data (Fig. 6).

First consider factor analysis (Fig. 6A), mentioned in the ordination techniques, in which there are many variables and the relationship between them is required from the data themselves. In the terminology of causal diagrams, this could be the case where there are unknown or unmeasureable environmental variables (x) whose effect we observe in the measured variables (y's) and the problem is to deduce the former from the latter. The method is to define a set of variables that are linear combinations of the measured variables and which account for the observed variation in the data. Since by definition the unknown variables are unmeasureable, the strategy adopted is to define the new variables so that the maximum effect is contributed by a first variable, then a second variable independent of the first, which accounts for most of the remaining variation, and so on. In practice one assumes that it will then be possible to give biological meaning to the variables so defined. The interpretation given here of regarding the measured variables as the dependent variables and variable defined by the method as independent variables is the converse of that usually given.

The more usual situation will be for the problem to progress through the stage of recognizing the necessary variables and the relationships between them, and for there to be measurements, perhaps first only qualitatively, then quantitatively. Thus the appropriate statistical technique is likely to progress through analysis of variance, covariance analysis to regression and path analysis.

In multiple regression, there are quantative measurements of all variables. The method determines the values and confidence intervals of the coefficients (b's) in the linear relationship between these variables in the sample data (Fig. 6B). The confidence intervals of the coefficients allow decision on whether variation of particular independent variables contributes significantly to the relationship at known levels of probability. Stepwise procedures allow independent variables to be included or excluded from the regression equation in order of significance. This is useful where only empirical predictive equations are required and one has to search among many variables for the group that show the highest correlation with the variable of interest.

In multiple regression (and almost all of the other techniques listed), it is assumed that the independent variables have been measured without error (unlikely) and that the only errors are the departures between the actual and estimated values of the dependent variables. The method differs slightly depending on the actual distribution of values of the variables and the departures (Table 2). In Type 1 or fixed regression models (and analysis of variance), the values of the independent variables are regarded as the only ones of interest, or fixed (and may have been chosen by the investigator), whereas in the Type II or random regression model, the particular values are regarded as a random sample of all possible levels of the variable. In Type I, the values of the original variables may be transformed, or new variables created which are functions of other variables. In relation to biological interpretation, the method assumes only a two stage relationship between the variables with the dependent variable in one hand and the independent variable(s) on the other.

Path analysis (TUKEY, 1954; FERRARI, 1963; SCOTT, 1966, 1973) is similar to multiple regression in determining the coefficients, confidence intervals, etc. of linear equations showing the relationships between variables for which there is a set of quantitative continuous data. The difference is its multistage considerations, in that a variable, which is a dependent variable in relation to one group of variables, may simultaneously be an independent variable in relation to an associated group of variables (Fig. 6C). The reason for this difference is that path analysis takes account of the errors in estimation in those dependent variables which appear as 'independent' variables in associated equations. It is this ability to handle multistage relationships which makes the method attractive in ecological problems.

The biological interpretation of linear equations is in terms of the coefficients giving the rates at which the values of plant variables change per unit change in the environmental variables.

Where the methods do not require that the independent variables are independently distributed then it is possible to create new independent variables which are a function of existing variables, *e.g.*, polynomial, logarithmic, or other transformations. By this means, curvilinear relationships between variables can be fitted in many circumstances and interpretation can be in terms of maxima, minima, or points of inflection. For example, PUTTER, *et al.* (1966) interpret a general quadratic relationship $y = b_0 + b_1 x_1 + b_2 x_2$ in the alternative form

$$y = A - B(x - C)^2$$

where C gives the optimum value of the environmental factor (x) at the maximum value A of the plant variable (y) and where B is regarded as a measure of how fast y decreases as x moves away from the optimum.

As Table 2 and the discussion show, the statistical requirements of the data for the various methods is often more restrictive than commonly believed. The only justification in practice is that many of the methods have been shown to give good approximation even when the underlying assumptions are only partly valid.

Even when some of the conditions are invalid, the mathematics of the techniques may still give the best estimates of some of the quantities sought (*e.g.*, the coefficients of the linear equation) even though it will not be possible to make any statements about their reliability. But many statistical parameters are of use only in the strict statistical sense (*e.g.*, correlation coefficient).

4.8 **Curvilinear Relationships**

Linear equation descriptions of processes involved in plant/environment relationships are likely to be the exception rather then the rule, and must be regarded as first approximations, or as approximations of segments of the full response surface. There is no general class of mathematical equations which are sufficiently flexible to give even empirical equations of the types of relationships expected (Fig. 5).

In practice, the elucidation and description of complex curvilinear relationships will depend on a combination of data accurate enough to provide leads or test alternative equations, and empirical or theoretical mathematical understanding of the processes involved. Only a few points will be made.

First, WATT (1968) has given some criteria for selecting the appropriate equation for the relationship between two variables based on the observed changes in gradient of graphs of the dependent variable (y) versus the independent variable (x). His four criteria: Are the changes in gradient:

1. direct or inversely linearly proportional to the independent variable?
2. linearly proportional to the dependent variable?
3. approaching zero as the dependent variable approaches an upper assymptote?
4. nearing infinite gradient as independent variable increases?

Listed below are the seven of the possible equations which WATT (1968) has found to most commonly occur in biological data.

$y = a + bX$ $\qquad$ $y = aX^b$
$y = a + b \ln X$ $\qquad$ $y = y_{max}/(1+e^{(c-bx)})$
$y = ae^{bx}$ $\qquad$ $y = y_{max}(1+e^{-bx})$
$y = a + bX - cX^2$

These equations show a second point. The first equation is already a linear relationship between the measured variables. The next four equations can be converted to linear form by appropriate transformation of the variables or equation (*e.g.*, the second equation by using the natural logarithms of the dependent variable) and thus can be fitted by linear multiple regression techniques providing the other necessary conditions are satisfied. Thus, the wide use of linear statistical methods has been a reflection of the investigator's ingenuity in making data amenable to linear techniques rather then of their occurrence in nature. The last two equations are among the large group of equations which cannot be converted to a linear form and have to be fitted by iterative methods (successive trial and error). WATT (1968) has also discussed some of the iterative techniques available.

An example of how consideration of the mechanism involved may lead to derivation of the possible mathematical form of the relationship between variables is VISSER's (1963) discussion of the relationship between yield (y) and several environmental factors (x's) considering three possible assumptions. These are listed for three environmental variables together with the corresponding equations in which A, b's, c's, and d are fitted coefficients (and in Fig. 7 for one variable).

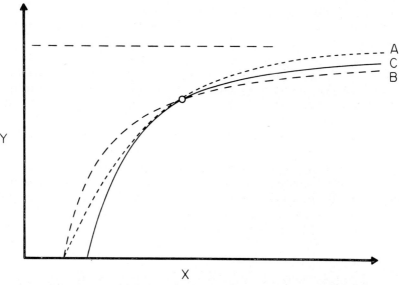

Figure 7. Comparison of Visser's three alternatives of relationship between yield (y) and a single environmental variable (x). Upper assymptote same and graphs coincident at three quarters of that value.

A. Yield increase proportional to deficit from optimum (Mitscherlich law of limiting factor)

$$y = A(1-e^{-(x_1-b_1)/c_1})(1-e^{-(x_2-b_2)/c_2})(1-e^{-(x_3-b_3)/c_3})(----)$$

B. Yield increase inversely related to level of environmental variable and proportional to deficit from optimum.

$$y = A \left(\frac{x_1-c_1}{x_1-b_1}\right)\left(\frac{x_2-c_2}{x_2-b_2}\right)\left(\frac{x_3-c_3}{x_3-b_3}\right)(-------)$$

C. Direct constant increase in yield with level of environmental variable and proportional to deficit from optimum.

$$(A-y)(b_1x_1-y)(b_2x_2-y)(b_3x_3-y)(----) = d$$

These show that even simple biological assumptions may lead to relatively complex equations and differing form of the relationship (Fig. 7). The same figure illustrates another point, that precise data would be required to discriminate between the alternatives.

Another type of problem involving curvilinear relationships lies in the separation of growth components involving a series of measurement over time. The growth of a plant in a period is related to its size and potential response at the beginning of the period and a variable increment related to fluctuations in environmental conditions during that interval (Fig. 1). The separation of these two components may be required.

Two related approaches have been used. Both use some form of logistic growth curve (CARMER & JACOBS, 1965; CARMER, 1964) to describe the variation in accumulated growth. In the first approach, the time scale is replaced by the time integration of environmental variables, *e.g.*, degree-days, accumulated evaporation, etc. — and where the relative closeness of fit for the different factors is taken as indicative of their relative importance (NELDER, *et al.* 1960; AUSTIN, *et al.* 1964).

The second approach is to take the deviations of the individual sampling points from the fitted logistic curve as being the environmental effect and to be tested by other methods. For example, BROUGHAM & GLENDAY (1969), and references therein) by suitable experimental design were able to separate growth and climate effects of pasture growth on a monthly, weekly, and daily basis.

There are several unresolved mathematical problems in relation to plant/environmental analysis. The difficulties of the selection and fitting of non-linear relationships has been mentioned. Also, the statistical bases of many of the curvilinear relationships are not completely understood so that often it will not be possible to make reliability statements about the fitted relationships. Also, most of the types of equations discussed have referred to the form of the response between the lower limit of the dependent variable and the optimum. None describe the full response relationship from the lower limit, to optimum and through to the upper limit. VISSER (1968) has discussed this problem.

Further, there appear to be no mathematical relationships which can adequately cope with data in which zero is a common state of some of the variables (Fig. 6A) (WILLIAMS & DALE, 1962). This is not as trite a problem as it may first seem. For example, any study along an environmental gradient will give many sites on which particular species is absent and, in many instances, it will be as useful to know the combinations of environmental conditions where the species does not occur, as to know its environmental relationships where it does occur.

In referring back to Fig. 1, it will be noted that there has been little discussion of techniques of defining the tolerance ranges of plants to environmental variables. There seem to be few appropriate mathematical techniques.

Little mention has been made of computers. Suffice to say that they are the basic means whereby biologists can readily use the techniques described. Examples of the systems approach and modelling will be discussed in other articles in this series.

The types of relationships discussed here may be included in a computer simulation of biological systems as conditional statements if they are only qualitative, as — a data matrix with interpolation procedures if they are in the tabular or graphical form, or as the appropriate equations, whether empirical or theoretical, where these have been established.

4.9 Conclusions

The relationship between plants or vegetation and environment is complex. I have tried to make the following points:

1. Because of the biological and physical process concerned, any single vegetation or environmental variable is only likely to have a direct effect on, or be directly affected by, a few other factors, and that complex multistage interactions are composed of these simpler relationships.

2. Causal diagrams can be used to show such relationships.

3. The knowledge of any particular interaction may progress from qualitative, through to graphical or tabular, and to empirical or theoretical quantitative equations.

4. That the mathematical techniques likely to be useful at each of these stages are association analysis, factor or component analysis, contingency tables, analysis of variance, multiple regression, and path analysis, to iterative curvilinear curve fitting.

5. Where probabilistic statements or confidence intervals are required then statistical methods are needed but that there are limitations to these techniques.

REFERENCES

Austin, R. B., J. A. Nelder & G. Berry – 1964 – The use of a mathematical model for the analysis of manurial and weather effects on the growth of carrots. *Ann. Bot.* 28: 153–162.

Billings, W. D. – 1952 – The environmental complex in relation to plant growth and distribution. *Quart. Rev. Biol.* 27: 251–265.

Billings, W. D. – 1974 – Environment: Concept and reality. Handbook of Vegetation Science (this volume).

Bray, J. R. & J. T. Curtis – 1957 – An ordination of the upland forest communities of southern Wisconsin. *Ecolog. Monog.* 27: 325–349.

BROUGHAM, R. W. & A. C. GLENDAY – 1969 – Weather fluctuations and the daily rate of growth of pure stands in three grass species. *N.Z.J. Agric. Res.* 12: 125–136.
CARMER, S. G. – 1964 – Expotential regression and a computer program for estimation of parameters. *Agron. J.* 56: 515.
CARMER, S. G. & J. A. JACOBS – 1965 – An expotential model for predicting optimum plant density and maximum corn yield. *Agron. J.* 57: 241–244
FERRARI, T. J. – 1963 – Causal soil-plant relationships and path coefficients. *Plant & Soil* 19: 81–96.
JACCARD, P. – 1932 – Die statistische – floristische Methode als Grundlage der Pflanzensoziologie. Abderhalden Handb. Arbeitsmeth. Abt. 11: 165–201.
LANCE, G. N. & W. T. WILLIAMS – 1965 – Computer programs for monothetic classification ("Association analysis"). *Computer J.* 8: 246–249.
LANCE, G. N. & W. T. WILLIAMS – 1967 – A general theory of classificatory strategies 1. Hierarchical systems. *Computer J.* 9: 373–380.
MAJOR, J. – 1951 – A functional, factorial approach to plant ecology. *Ecology* 32: 392–412.
MASON, H. L. & J. H. LANGENHEIM – 1957 – Language analysis and the concept Environment. *Ecology* 38: 325–340.
NELDER, J. A., R. B. AUSTIN, J. K. A. BLEASDALE & P. J. SALTER – 1960 – An approach to the study of yearly and other variation in crop yield. *J. Hort. Sci.* 35: 73–82.
PUTTER, J., D. YARON, & H. BIELORAI. 1966; Quadratic equations as an interpretative tool in biological research. *Agron. J.* 58: 103–104.
RANDALL, J. M. – 1969 – An introduction to association analysis. *Proc. N.Z. Ecol. Soc.* 16: 48–57.
SALISBURY, F. B., G. G. SPOMER, M. SOBRAL & R. T. WARD – 1968 – Analysis of an alpine environment. *Bot. Gaz.* 129: 16–32.
SCOTT, D. – 1966 – Interpretation of ecological data by path analysis. *Proc. N.Z. Ecol. Soc.* 13: 1–4.
SCOTT, D. – 1969 – Relationships between some statistical methods. *Proc. N.Z. Ecol. Soc.* 16: 58–64.
SCOTT, D. – 1973 – Path analysis: A statistical method suited to ecological data. *Proc. N.Z. Ecol. Soc.* 20: 79–95.
SCOTT, D. & W. D. BILLINGS – 1964 – Effect of environmental factors on the standing crop and productivity of an alpine tundra. *Ecol. Monog.* 34: 243–270.
TRACEY, J. G. – 1969 – Edaphic differentiation of some forest types in eastern Australia. I. Soil physical factors. *J. Ecol.* 57: 805–846.
TUKEY, J. W. – 1954 – Causation, regression and path analysis p. 35–66. *In* O. KEMPTHORNE, T. A. BANCROFT, J. W. GOWEN, & J. T. LUSH (eds.) Statistics and Mathematics in Biology. Iowa State College Press (Ames) 632 pp.
VISSER, W. C. – 1963 – Formulae for the ecological reactions of crop yields pp. 326–339 *in* "The Water Relations of Plants". Blackwell 1963.
VISSER, W. C. – 1969 – Mathematical models in soil productivity studies, exemplified by the response to nitrogen. *Plant & Soil* 30: 161–182.
WATT, K. E. F. – 1968 – Ecology and Resource Management. McGraw-Hill New York 450 pp.
WILLIAMS, W. T. & M. B. DALE – 1962 – Partition correlation matrices for heterogeneous quantitative data. *Nature* 196: 602.

5 ALLELOPATHY IN THE ENVIRONMENTAL COMPLEX

Cornelius H. Muller

Contents

5.1	Introduction	73
5.2	Relation to Competition	74
5.3	Function of Allelopathy in Dominance	76
5.4	Role of Dominance in Stability	81
5.5	The Limits of Allelopathy	82

5 ALLELOPATHY IN THE ENVIRONMENTAL COMPLEX[1]

5.1 Introduction

The biochemical parameters of the biosphere constitute a set of ubiquitous factors which, like temperature, may be high or low but never absent. Their qualities, in addition to intensity of concentration, may include capacity for stimulation or suppression of plant growth or they may simply be benign. When the biochemical factors suppress any plant growth, the process is called allelopathy (MOLISCH, 1937). The chemical products of a plant species are often highly selective in their inhibitory action, suppressing some species but not others. The toxic species may suffer autointoxication, as in some species of *Helianthus* (CURTIS & COTTAM, 1950), or it may be relatively tolerant of its own metabolic effluence.

In order for allelopathy to occur, there must exist a mechanism whereby toxins are released into the environment, concentrated and retained at a site of action, and absorbed by the victim species. Lacking any essential step in this mechanism, phytotoxic growth suppression cannot develop regardless of existing toxicity within the plant.

Interaction between toxicity and other natural stresses often produce synergistic effects so that the limiting quality of phytotoxicity itself may be subtle or obscured. Indeed, there exist relative to phytotoxicity as many instances in which it is not a limiting factor as may be ascribed to any other factor in the environmental complex. Conversely, however, limiting status of the biochemical factor has been established in a wide variety of situations.

Phytotoxicity plays a role of importance in most of the traditionally recognized processes in plant ecology (MULLER, 1969). A consideration of a few of these will suffice to illustrate the functioning of allelopathy within the environmental complex.

---

[1] Acknowledgement is due the National Science Foundation for financial support over a decade of my research and that of my students from which this is an outgrowth. The Santa Barbara Botanic Garden has furnished a quiet refugium for writing.

5.2 Relation to Competition

The deleterious effect of one plant upon another in the broad sense has been very aptly termed "interference" by HARPER (1961) in order to avoid precisely the kind of specificity of meaning which now becomes necessary. Because these deleterious effects may take the form of depletion of some necessary substance or form of energy existing in limited supply, such instances constitute "competition" in its original sense (as defined, *e.g.*, by OOSTING, 1956). This does not embrace phytotoxicity which involves the addition of an unfavorable substance and no depletion at all. In fact, as a consequence of toxic suppression and elimination of some plants, the levels of such factors as light intensity and soil moisture may increase. This non-depleting interference is "allelopathy". The importance of making this distinction, which has long been practiced among European authors (*e.g.*, BÖRNER & RADEMACHER, 1957), lies in the necessity of accounting for the one sort of interference in order to validate the significance of the other. The close proximity upon which interference is dependent renders both competition and allelopathy possible until experimentation demonstrates one to be operative and the other not. The wide usage of "competition" to embrace all interference has therefore introduced a serious semantic problem for experimental synecology.

Studies of competition have long been generally devoid of consideration of the possibility of phytotoxic involvement. If two plants are in such close proximity that they draw upon the same supplies of water and mineral nutrients, they are also close enough to be immersed in, or touched by, one another's biochemical products. Experimental designs that fail to separate these two classes of interfering reactions are not reliable means of demonstrating either competition or allelopathy. As allelopathy is recognized in growing numbers of vegetational situations, it becomes increasingly obvious that many of the early studies of competition for minerals, soil moisture, and even light will need to be repeated or re-evaluated.

It is possible, and indeed probable, that competition and allelopathy may share limiting significance in some situations. This does not mean, however, that they are not individually limiting in other situations. For example, there exist shade tolerant plants which show variously restricted distributions within mixed forests on uniform soil and topography as well as forest dominants of greatly divergent toxic potential. Thus, one is justified in seeking limiting biotic factors which show distinct control. An excellent case was described by HOOK & STUBBS (1967) in which isolated

Figure 1. Differential undergrowth associated with isolated seed trees in a bottomland forest in South Carolina 4 years after cutting: A, *Liquidambar styraciflua* with normal growth of briers, shrubs, and tree samplings; B, *Quercus falcata* var. *pagodaefolia* with few retarded tree seedlings and a grassy ground cover. (From HOOK & STUBBS, 1967; courtesy of the Southeastern Forest Experiment Station, U. S. Forest Service.)

seed trees left after cutting of a lowland mixed hardwood forest in South Carolina showed strong divergence in their interference with the regeneration of woody species (Fig. 1). DeBell (1971), in an experimental study of this phenomenon at the same site, established the sufficiency of all requisite physical factors and the biochemical quality of the interference. He showed that soil from beneath *Quercus falcata* var. *pagodaefolia* inhibited seedling growth of *Liquidambar styraciflua* in both field and pot experiments much more than that of *Quercus* seedlings. In the converse situation, *Liquidambar* soil failed to show capacity for any seedling inhibition. DeBell furthermore demonstrated toxicity of aqueous leachate from *Quercus* leaves and identified the phytotoxin as salicyclic acid. Prior to the initial cutting of this forest, light reduction might well have limited the growth of plants tolerant to toxins. After cutting and light increase to full sunlight, toxic chemicals continued to limit the growth of plants of most woody species at specific sites.

There is a potential negative relation between allelopathy and competition. If a species has the capacity to compete with another for some habitat factor in short supply, close proximity of the two will result in competition. However, should one of these have the allelopathic potential to exclude the other completely, then competition never develops. One cannot cite examples of this because plants must grow together in order to be tested for competition. Who knows how many mutually exclusive species pairs might compete if grown to maturity in mixed stands? Allelopathy, on the other hand, can be demonstrated between mature plants of one species and seedlings of the other if all physical factors are kept optimal. There emerges, as a consequence of allelopathy, an avoidance of competition, if one accepts the competitive potential between such a pair of allopatric species.

5.3 Function of Allelopathy in Dominance

In the functional sense, dominance may be defined as a reaction of plants upon the environment sufficiently strong to influence which other plants will grow there. The reaction is likely to constitute interference and to result in the exclusion of all except tolerant species. Whether the interference involved is competitive or allelopathic, the resulting exclusion depends upon the intensity of the reaction and the degrees of susceptibility exhibited by other species. Thus, a forest tree may dominate by casting dense shade within which only shade-tolerant species will grow. Or a phytotoxic dominant, such as *Quercus falcata* var. *pagodaefolia* or *Juglans nigra* may exclude all but a few tolerant understory species.

Because exclusion is a common expression of dominance, one frequently faces obscure examples. This is particularly true of allelopathic exclusion, which is often complete. One then sees only the final result, the absence of the excluded species. It is understandably difficult to deal with a cryptic concept such as the exclusion of the "invader" that is not there. The importance of this is evident in the great residue of unresolved differences of opinion centering about the factors controlling invasion of grasslands by woody plants and similar problems related to other homeostatic vegetations. The more efficient the mechanism of exclusion, the greater the difficulty of seeing the process in action.

Dominance has a significant time dimension (BECKING, 1969) within which reaction and response occur. Duration of reaction determines the periods during which susceptible species will encounter intolerable conditions, whether phytotoxic or competitive. Longevity of dominants, periodicity of their reaction activities, and persistence of their toxic effluents are among the variables that determine the time dimension of dominance. In instances involving synergistic relations between phytotoxins and other environmental stresses, additional variables may become important. For example, the toxic action of *Salvia leucophylla* depends upon normal, low-level drought stress. A period of several years of favorable soil moisture resulting from an increase in precipitation effectiveness caused a significant reduction in dominance of shrubs over annual grasses (MULLER, 1970). During the same period, neighboring chaparral dominated by *Adenostoma fasciculatum* and *Arctostaphylos glandulosa* var. *zacaensis*, having no such synergistic dependence, exhibited no relaxation of its phytotoxic dominance over excluded herbs. A strong but irregular periodicity results from recurrent fires which interrupt the production of toxins and apparently denature those already in the environment. This is vividly illustrated by the fire cycle of California chaparral (MULLER, HANAWALT, & MCPHERSON, 1968) in which fire is followed by a flush of herb and shrub seed germination and luxuriant growth, all completely suppressed by chemical inhibition during periods of shrub dominance prior to fire. Another tendency toward toxic periodicity is apparent among deciduous dominants whose maximal toxicity occurs in fallen leaves. Under climatic conditions favoring rapid decomposition of leaf litter, renewed toxicity occurs with each leaf fall. Thus, MCPHERSON (1971) has shown allelopathic effects of leaf litter in mixed stands of *Quercus stellata* and *Q. marilandica* invading prairie with annual precipitation in excess of 825 mm and consequent prompt litter decomposition. He found that periodic removal of oak leaf litter resulted in about a 10-fold increase in seed germination of characteristic understory species.

The best examples of strong dominance may well be those situations in which plants form extensive single-species stands. To be sure, some extreme environments may select one or a few species so that dominance may thus become more apparent than real. In the absence of such selection, however, purity of stand is a reflection of exclusion of other species by the dominant.

In each instance of single-species dominance, it is necessary to examine the process of exclusion very closely before ascribing to it a competitive or an allelopathic cause. There exist many times more examples of pure stands than the few that have been subjected to satisfactory analysis. Many of these exhibit characteristics highly suggestive of allelopathy and would make excellent subjects of study. For instance, both *Fagus grandifolia* and *Acer saccharum* sometimes form pure stands.

Trees of the two species cast similar shade and have equally strong concentration of absorbing roots in the upper soil horizons. Although trees of both species tend to harbor fewer understory plants than do those of numerous deciduous species with which they intermix elsewhere, *Fagus* is much the more strongly inclined to inhibit undergrowth. Neither has been examined critically for interference mechanisms. Numerous coniferous trees occur as pure stands. While some such forests may be the consequence of stringent habitat selection, others appear to be excluding neighboring species from apparently suitable sites. *Tsuga canadensis* has a strong shading reaction as well as an abundant discharge of phenolic compounds which appear to be allelopathically effective. Experimental distinction between these reactions is still lacking. The list of as yet incompletely or totally unstudied pure stands could be greatly extended by including single-species grass sods, oak shrub thickets, various shrubs of the American Great Basin and similar cold deserts, and multitudes of others.

Among the more completely studied examples of single-species dominance are *Adenostoma fasciculatum* and *Arctostaphylos glandulosa* var. *zacaensis* which are involved in the fire cycle of California chaparral. In the instance of *Adenostoma*, all pertinent physical and biotic factors were examined (MCPHERSON & MULLER, 1969) in an effort to determine the cause of inhibition of germination and growth among suppressed associated species. Open canopies permitting full sunlight, upper soil horizons devoid of shrub roots, uniform distribution of mineral nutrients in time and space, uniform soil moisture distribution during periods of herb growth, and failure of herbs to respond to supplemental mineral applications effectively ruled out physical factors and competition. On the other hand, experimental soil heating resulted in a distinct flush of seed germination and growth of seedlings on the bare soil beneath and between shrub canopies. This, together with demonstration by bioassays of a definite toxicity of rain drip from foliar crowns, strongly suggested the existence of heat-labile inhibitory chemicals in the system. Removal of foliar crowns by clipping at soil level resulted, during the first growing season thereafter, in a great flush of herb and shrub seedlings, luxuriant growth and maturation of annual herbs, and species composition entirely comparable to growth response after fire described by SAMPSON (1944) and SWEENEY (1956). Detailed analysis of rain drip from *Adenostoma* canopies (MCPHERSON, CHOU & MULLER, 1971) revealed nine phenolic acids, quinones, and coumarins, most of which were highly toxic in bioassays. Seven of these, furthermore, were isolated from soil beneath *Adenostoma* shrubs. This instance of dominance is thus clearly indicated to be phytotoxic in nature.

An even stronger allelopathic dominance was exhibited by *Arctostaphylos glandulosa* var. *zacaensis* (CHOU & MULLER, 1972). Pure stands of this shrub suppressed herbs beneath their canopies after the manner of *Adenostoma*. They also exhibited uniquely greater toxic potency and longer persistence of chemical inhibitors in the environment than did *Adenostoma*. Experimental clearings failed to show herb growth in the first growing season after clearing and made only a weak beginning in the second growing season. Old clearings from which shrubs were totally removed along with leaf litter and root crowns may or may not produce abundant herb growth. If such a clearing is topographically situated so that it receives no surface run-off from adjacent *Arctostaphylos* shrubs, the growth of annual herbs continues year after year. But if run-off from shrub thickets reaches the clearing, in a period of less than 10 years germination ceases and bare soil results. Such dominance has been observed 3 to 4 m distant from the shrubs (Fig. 2).

Arctostaphylos releases into the environment numerous phenolic acids, quinones, and coumarins most of which are potent phytotoxins. They have been isolated in quantity from foliar rain drip and especially from leaf litter. Transformation of these compounds in soil is apparently fairly rapid, for mineral soil beneath the shrubs yielded principally other (apparently derivative) phenolic acids and coumarins, all of which were significantly phytotoxic. The role of microorganisms in this system is little known, but if they are a significant toxin source, the specific nature of the microbial role appears to be uniquely associated with *Arctostaphylos*.

Figure 2. An old clearing on which annual herb growth was observed to decline and cease over a period of a few years; the area is under the biochemical control of *Arctostaphylos glandulosa* var. *zacaensis* from which run-off flows. © Copyright by C. H. MULLER, 1972.

5.4 **Role of Dominance in Stability**

The condition of stability ascribed to vegetation (as in the concept of "climax") is actually a homeostatic phenomenon, involving a limitation upon flux. The minor fluctuations of physical factors, the small disturbances, mortalities, and invasions in even the most stable systems would open the way to changes of successional magnitude were there not some limitations inherent in vegetational organization. In some extraordinary environments, such as desert and tundra, (SHREVE 1942; MULLER 1941, 1951) these limitations lie in physical extremes which reduce the number of species capable of ecesis and thereby may eliminate conventional succession. In the resulting "auto-succession", the original dominants initiate revegetation of disturbed areas without the intervention of other species. Thus, in drastically disturbed extreme desert and arctic tundra, both the first and the last stages of revegetation consist of the same species and, indeed, the same individual plants.

Toward the favorable end of the spectrum of environmental rigor, the lack of extreme factors leaves the stability of vegetation dependent upon dominance. The ability of dominant species to select associates by excluding others, if sustained by longevity, leads to an effective stability so long as the dominants are not destroyed by some form of disturbance and produce seedlings tolerant of the mature vegetation. Because dominance has traditionally been regarded as operating through control of physical factors (soil genesis, light reduction, mineral control, and mitigation of moisture extremes), the equally important role of biochemical factors has been widely disregarded.

The favorable aspects of the biochemical parameters, if not understood, are at least widely recognized as the basis of the "living" soil system. Decomposition, recycling, a multitude of symbioses, the tilth, and the stimulatory qualities of this system are dependent upon the exudation, leaching, exfoliation, and ultimate mortality of plants. While this recognition is implicit in the concept of "climax" in forests and grasslands, no provisions have been made in most discussions of this concept for the biochemical inhibitory qualities of dominant plants. Even successional concepts have depended upon the idea that one stage renders the habitat more favorable for the next stage and thus permits invasion by species previously excluded by hostile physical soil conditions. Rice and his students (*e.g.*, RICE, 1968; WILSON & RICE, 1968; PARENTI & RICE, 1969; OLMSTED & RICE, 1970) have offered strong evidence that succession of abandoned fields in Oklahoma includes several allelopathic relations, including removal of some pioneer species by autointoxication and persistence of dominants favored by low nitrogen through their interference with nitrogen-fixing and nitrifying organisms. There is good reason to speculate that the dominants of the final stage are also allelopathic. There is needed a study to determine whether or not the homeostatic condition of prairie is as much dependent upon phytotoxic as upon competitive exclusion of the weedy pioneer species. There is no need for speculation, however, in the cases of many dominants of homeostatic vegetations that have been investigated for phytotoxic dominance. *Adenostoma*, *Arctostaphylos*, *Juglans*, several species of *Quercus*, *Juniperus*, and others that have been critically studied have shown significant allelopathic activity.

The significance of homeostasis in successional stages short of "climax" looms large in the functional analysis of ecosystems. The ecosystem sometimes exhibits abrupt deviation from the expected pattern of productivity and nutrient cycling. Strong toxic potential permits dominance involving extraordinarily low levels of biomass. Thus, areas within grassland, heath, and even montane tropical forest are held for indefinitely long periods by pure stands of *Pteridium aquilinum* (GLIESSMAN & MULLER, 1972). Every instance of long tenure in a secondary successional series by species characterized by relatively small, weak plants therefore becomes suspect and requires evaluation of phytotoxic involvement. Toxin production at effective levels simply requires less biomass than does effective shade or competition for water. The determining parameters of the ecosystem therefore consist not solely of the quantity of resources present but may also include the negative action of otherwise unimpressive toxic plants.

Toxic dominants subject to autointoxication may induce instability. The intensity of phytotoxic potential, the susceptibility of the dominants (and their progeny) to specific toxins, and the longevity of the dominants all play against the time factor to produce a rate of change. Thus degrees of instability (as well as stability) may issue from a single widespread and variable mechanism.

5.5 The Limits of Allelopathy

While a great preponderance of plant species can be shown to contain potent phytotoxins, so that one is tempted to regard allelopathy as a universal attribute of vegetation, this conclusion would be a disastrous error. It is generally true that rigorous environments with scanty plant cover seldom permit sufficient concentration of phytotoxins to cause perceptible allelopathy. Thus, in deserts there has not been established a single authentic instance, although several studies have purported to do so. Such efforts have usually not given due consideration to the superior limiting qualities of extreme physical factors and have invariably failed to establish a mechanism whereby phytotoxins are released, concentrated, and absorbed. It would be difficult to overemphasize the caution required in order to avoid unfounded conclusions.

The necessary mechanism of allelopathy has a definite set of critical requirements which may be met in different ways by different plants in various situations. These will be presented seriatum with some illustrative examples of failures capable of destroying allelopathic effectiveness and some oversights that could lead to logical error.

1. The plant must produce a toxin. This need not be noxious to the plant in which it occurs (although it may have that potential) and may actually occur in detoxified condition, such as most of the phenolic glycosides. Some plants reveal no toxicity in bioassays and appear to be totally benign toward their associates.

2. There must exist a means of release of toxins into the environment. Aromatic species evaporate terpenes into the atmosphere, a process enhanced by arid and semi-arid climates. Others are leached of the water-soluble toxins in their foliage by rain and fog, a means made possible by frequent precipitation. Water-soluble toxins of plants confined to dry areas accumulate in fruits, senescent leaves, and other organs or tissues being cast off.

3. A means of movement and concentration of toxins must exist. Terpenes and phenolic acids are adsorbed upon colloidal soil constituents. In the instance of *Eucalyptus camaldulensis* the effectiveness of toxic phenolic acids was found to be dependent upon a clay-loam soil (DEL MORAL & MULLER, 1970); trees growing on sand failed to inhibit herbs that were fully inhibited on heavier soil. Confirmation of toxin presence in soil by extraction and identification is highly desirable.

4. Susceptibility of apparently inhibited plants must be established. In a sense, bioassays constitute indirect evidence because they require higher concentrations of toxic material than occurs in nature in order to overcome the unusually favorable growing conditions characteristic of *in vitro* bioassay designs. Supplementary greenhouse and field experiments are often more convincing. Wholly tolerant plants effectively circumvent allelopathy.

5. Elimination by controlled experimentation of physical factors and non-chemical biotic factors is completely mandatory. Such limiting factors may obscure the lack of one or several steps in

a presumed allelopathic mechanism.

Each local instance of the failure of a known allelopathic plant to exhibit effective inhibition is clear evidence that allelopathy is no panacea and that it may not be regarded as ecologically universal. Additionally, those dominants that fail, for lack of some essential step in the mechanism outlined above, to show evidence of biochemical dominance are further evidence of less than universal significance of allelopathy.

The biochemical parameter set is justifiably regarded as one of the major classes of environmental factors, such as light, temperature, moisture, and mineral nutrients. All of these, biochemicals included, are participatory wherever plants grow and individually limiting here and there in the biosphere.

REFERENCES

BECKING, R. W. – 1968 – Vegetational response to change in environment and change in species tolerance with time. *Vegetatio* 16: 135–158.

BÖRNER, H. & B. RADEMACHER – 1957 – Untersuchungen zum Problem der echten selbstunverträglichkeit des Leins (*Linum usitatissimum* L.). *Zeitschr. Pflanzenerähr., Dung., Bodenkunde* 76 (121): 123–132.

CHOU, C.-H. & C. H. MULLER – 1972 – Allelopathic mechanisms of *Arctostaphylos glandulosa* var. *zacaensis*. *Am. Midl. Nat.* 88: 329–347.

CURTIS, J. T. & G. COTTAM – 1950 – Antibiotic and autotoxic effects in prairie sunflower. *Bull. Torrey Bot. Club* 77: 189–191.

DEBELL, D. S. – 1971 – Phytotoxic effects of cherrybark oak. *Forest Science* 17: 180–185.

DEL MORAL, R. & C. H. MULLER – 1970 – The allelopathic effects of *Eucalyptus camaldulensis*. *Amer. Midl. Nat.* 83: 254–282.

GLIESSMAN, S. R. & C. H. MULLER – 1971 – The phytotoxic potential of bracken, *Pteridium aquilinum* (L.) Kuhn. *Madroño* 21: 299–304.

HARPER, J. L. – 1961 – Approaches to the study of plant competition. In Mechanisms in Biological Competition, ed. MILTHORPE, F. L. *Symposia of the Society for Experimental Biology* 15: 1–39.

HOOK, D. D. & J. STUBBS – 1967 – An observation of understory growth retardation under three species of oaks. *U. S. Forest Service Res. Note SE-70.* 7 p.

MCPHERSON, J. K. – 1971 – Interference of *Quercus marilandica* and *Q. stellata* with understory plants (Personal communication).

MCPHERSON, J. K. & C. H. MULLER – 1969 – Allelopathic effects of *Adenostoma fasciculatum*, "chamise", in the California chaparral. *Ecol. Monogr.* 39: 177–198.

MCPHERSON, J. K., C.-H. CHOU & C. H. MULLER – 1971 – Allelopathic constituents of the chaparral shrub *Adenostoma fasciculatum*. *Phytochem.* 10: 2925–2933.

MOLISCH, H. – 1937 – Die Einfluss einer Pflanze auf die andere – Allelopathie. Jena. 106 p.

MULLER, C. H. – 1940 – Plant succession in the *Larrea-Flourensia* climax. *Ecology* 21: 206–212.

MULLER, C. H. – 1952 – Plant succession in arctic heath and tundra. *Bull. Torrey Bot. Club* 79: 296–309.

Muller, C. H. – 1969 – Allelopathy as a factor in ecological process. *Vegetatio* 18: 348–357.
Muller, C. H. – 1970 – The role of allelopathy in the evolution of vegetation. *In* Biochemical Coevolution, ed. Chambers, K. L. *Proc. 29th Annual Biol. Colloq., Oregon State Univ.*, 1968: 13–31.
Muller, C. H., R. B. Hanawalt & J. K. McPherson – 1968 – Allelopathic control of herb growth in the fire cycle of California chaparral. *Bull. Torrey Bot. Club* 95: 225–231.
Olmsted, C. E. & E. L. Rice – 1970 – Relative effects of known plant inhibitors on species from first two stages of old-field succession. *Southwest. Nat.* 15: 165–173.
Oosting, H. J. – 1956 – The Study of Plant Communities. 2nd ed. San Francisco. 440 p.
Parenti, R. L. & E. L. Rice – 1969 – Inhibitional effects of *Digitaria sanguinalis* and possible role in old-field succession. *Bull. Torrey Bot. Club* 96: 70–78.
Rice, E. L. – 1968 – Inhibition of nodulation of inoculated legumes by pioneer plant species from abandoned fields. *Bull. Torrey Bot. Club* 95: 346–358.
Shreve, F. – 1942 – The desert vegetation of North America. *Bot. Rev.* 8: 195–246.
Sampson, A. W. – 1944 – Plant succession on burned chaparral lands in northern California. *Univ. Calif. Agr. Exp. Sta. Bull.* 685. 144 p.
Sweeney, J. R. – 1956 – Responses of vegetation to fire. *Univ. Calif. Pub. Bot.* 28: 143–250.
Wilson, R. E. & E. L. Rice – 1968 – Allelopathy as expressed by *Helianthus annuus* and its role in old-field succession. *Bull. Torrey Bot. Club* 95: 432–448.

6 CORRELATION OF VEGETATION WITH ENVIRONMENT: A TEST OF THE CONTINUUM AND COMMUNITY - TYPE HYPOTHESES

Jon T. Scott

Contents

6.1	Introduction	89
6.2	Is Vegetation Quantifiable?	90
6.2.1	A Hypothetical Experiment	90
6.2.2	The Plant Community Hypotheses	92
6.2.3	Environmental Factors	96
6.2.4	Succession, Site Potential and Environmental Factors	97
6.2.5	Space-Related Factors	99
6.2.6	Environmental Synthesis	102
6.3	The Contest	104
6.4	Perspective	106

6 CORRELATION OF VEGETATION WITH ENVIRONMENT: A TEST OF THE CONTINUUM AND COMMUNITY - TYPE HYPOTHESES

6.1 Introduction

The science of vegetation deals with an extremely complex subject matter. Vegetation varies in time and space and is changed by a variety of environmental conditions including chance events and many kinds of disturbances. As discussed earlier in this part (BILLINGS, D. SCOTT) environment is composed of many elements each of which varies in time and space. It is no wonder, then, that ecologists striving for a simply expressed theory of vegetation-environment relations have not yet been successful. A fully developed, fruitful, economical and simple theory which explains all of the evidence has not yet emerged.

This section deals with the analysis of vegetation along environmental gradients. It is not intended to review the subject completely because this has been done in other chapters of the Handbook. The purpose here is to review those concepts and methods which, in the author's opinion, specifically apply to the analysis of vegetation-environment interactions.

Two generally accepted hypotheses on the relation between vegetation and environment have emerged which need further testing. To simplify the discussion here these shall be termed the "plant community hypotheses" and are labeled the *community-type hypothesis* and the *continuum hypothesis*. The purpose of this section is to provide a rigorous statement of these hypotheses and to propose methods which can be used to investigate further the long standing controversy (*c.f.*, DAUBENMIRE, 1966; MCINTOSH, 1967; DANSEREAU, et al., 1968; LANGFORD & BUELL, 1969).

In discussing the nature of vegetation many authors tend to choose a line of reasoning which favors one or the other of the two hypotheses. Others, however, have expressed the opinion that vegetation has characteristics of both. The attempt will be made here to define the problem so that only one hypothesis can be correct. That is, the two will be mutually exclusive. However, it will also be shown that elements of each hypothesis can be correct utilizing a relaxed definition.

The analysis of vegetation-environment relationships parallels the development of techniques of vegetation description, *i.e.*, "classification" (LAMBERT & DALE, 1964) and "ordination" (GOFF & COTTAM, 1967). Either method can be employed to gain further insight into the understanding of vegetation. A resolution of the *true* nature of the relation between vegetation and environment must utilize quantitative procedures which require no prior assumption of the nature of the relation. Some vegetation classification schemes have not been so rigorously achieved. Although they may provide utility they must be rejected for purposes of hypothesis testing.

While the testing of plant community hypotheses does not depend upon whether classification or ordination methods are used, this section will deal more with the use of techniques of ordination since they are more easily adapted to the study of the vegetation-environment relation. A proper test of the plant community hypotheses can be designed with classification methods if so desired. No attempt has been made to provide a complete literature review because of the extensive review in the preceding volume edited by R. H. WHITTAKER.

6.2 Is Vegetation Quantifiable?

6.2.1 A HYPOTHETICAL EXPERIMENT

Let us first deal with the concepts of vegetation and environment in relatively simple terms in order to plan a suitable experiment for hypothesis testing. Allow that a suitable measure of each can be obtained for an area on the earth which can be called a "stand". The vegetation will be measured to obtain a quantitative expression of the "importance" of each species in the stand. The important variables of the environment, such as properties of the climate and soil, will be measured.

The measures of vegetation and environment may be expressed on more than one axis (*e.g.*, multi-dimension ordination) but for simplicity we will consider only one dimension of each of these coordinates. These measures of a stand will be termed the *vegetation* and *environmental* indices. Each of the indices must be linear or somehow linearized. That is, a given magnitude of variation in either composition or environment must be equally expressed through the entire range of each index. Examples of how to obtain a vegetation (or compositional) index may be found in many studies including CURTIS & MCINTOSH (1951), BRAY & CURTIS (1957), AUSTIN & ORLOCI (1966), and ORLOCI (1966). The subject is reviewed by GREIG-SMITH (1964), WHITTAKER (1967), MC-INTOSH (1967), and GOFF & COTTAM (1967).

There are some prior conditions which must be set on the type of environmental measurement to allow in our experiment. Since it is known that environment may vary in a complicated way in space we must not include space-related variables such as elevation, slope aspect, or distance along the ground relative to an abrupt change in environment much as a lakeshore or swamp. The environmental measures should also not be affected by the vegetation itself and in fact should be independent of the vegetation. This last criterion is difficult to define and even more difficult to maintain in a field experiment. It will be discussed in more detail later in this section.

It is difficult to find studies in the literature which have maintained the rather restricted definition of environment proposed here. Most of the pertinent studies have either not determined an index value or have used measures which are not strictly independent of space along the ground (WHITTAKER (1956, 1960), MAYCOCK & CURTIS (1960), AYYAD & DIX (1964), WARING & MAJOR (1964), and WHITTAKER & NIERING (1964)). From the standpoint of the conditions specified here, the most satisfactory studies have been by LOUCKS (1962) and MONK (1965). The subject is reviewed by GREIG-SMITH (1964), MCINTOSH (1963, 1967), and WHITTAKER (1967).

Two additional criteria for testing the plant association hypotheses are that the samples (or stands) be reasonably stable and that they are spatially uniform. That is, the composition of the vegetation should not be changing significantly in time and should be reasonably homogeneous spatially. It would be more rigorous to require that the environment of the stand be spatially and temporally uniform but this is more difficult for definition and measurement.

Finally, our hypothetical test of the plant community hypotheses will require that a number of stands be sampled along a rather large range of environmental conditions or stated similarly over a large range of species composition. The sample need not be random provided we have kept the condition that the environmental index include only measurements that are truly independent of the vegetation and of space. To set the most rigid conditions on our test we should either have a very large number of samples or sample randomly along the environmental gradient. An alternative would be to determine the mean of an equal and large number of samples in each of several sub-ranges of the total range of environment. In this case we will not be able to test the reliability of means as we would in the case of the random sample.

The criteria for testing the continuum and community-type hypotheses can be summarized as follows: (1) stand (or sample) index values of vegetation composition and environment must be obtained objectively and quantitatively and presented on a linear scale; (2) the environmental measurements used must be independent of space and of the vegetation itself; (3) the vegetation (and preferably also the environment) within each sample must be reasonably uniform in space and time (i.e., homogenous and reasonably stable); and (4) the vegetation samples must be taken over a large range of environments, preferably on a random basis with respect to the environment.

6.2.2　THE PLANT COMMUNITY HYPOTHESES

Once the above criteria are met, we can state the plant community hypotheses in terms of vegetation-environment coordinates. Four hypothetical cases are given in Figure 1. The vegetation index is plotted on the ordinate as the dependent variable and the independent variable is the environmental index on the abscissa. The lines or graphs can be obtained by plotting mean values over ranges of the environmental index or by curve-fitting. A perfect fit of the curve in such a study would not be expected since there would no doubt be some degree of scatter in the original plot caused by chance factors or errors in measurement. Although the model is essentially stochastic, the analogous analytic function is expressed in Figure 1.

There are four hypothetical cases to be discussed once the proper statistical relation is determined. If the result is linear as in case (a) the continuum hypothesis would be correct for the region of the study. The vegetation gradient (or compositional gradient) is expressed by the slope of the curve. The vegetation continuum is defined as a constant vegetation gradient ($\Delta V/\Delta E$ = constant).

If the slope of the curve in Figure 1 is not constant then there must be some cause for "association" or clumping of species. Case (b) expresses gradual change of vegetation with respect to environment followed by more rapid change. In this case, neither the continuum nor community-type hypotheses are strictly defined and some "middle ground" is the actual result. Since the slope $\Delta V/\Delta E$ is not constant, the continuum hypothesis is not correct for this case. The community-type hypothesis is also not rigorously defined in case (b) since a "type" or "class" of vegetation would be most rigorously defined as a region of environment where V =

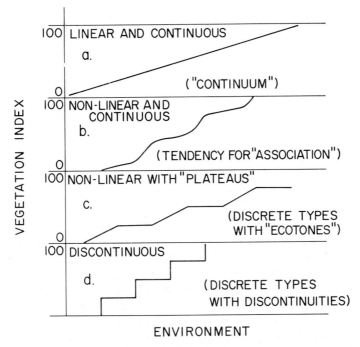

Figure 1. Hypothetical results of plotting a community vegetation index against a measure of community environment. See text for discussion of the four cases.

constant (or $\Delta V/\Delta E = 0$).

The community-type hypothesis can be stated in two ways as shown in cases (c) and (d). In case (c), the community-types (or "associations") where $\Delta V/\Delta E = 0$ are bordered by ecotones where the vegetation gradient is steep ($\Delta V/\Delta E$ is large). In case (d) the community-types are bordered by discontinuities where $\Delta V/\Delta E$ is not defined. Thus, case (a) defines the continuum hypothesis and cases (c) and (d) define two forms of the community-type hypothesis with case (d) being the most restrictive. Case (b) defines neither hypothesis but expresses characteristics of each.

The two plant community hypotheses have now been defined in a simple analytic coordinate system. Although the task of testing these hypotheses is not simple, because the required measurements are difficult, it is my opinion that the criteria must be adhered to and the vegetation-environment coordinate system must be used.

A large amount of work has been published utilizing the concept of species individuality following the early ideas of H. A. GLEASON (1939) and others. WHITTAKER (1967) terms this approach direct gradient analysis and discusses its use in the preceding volume of this handbook. This method utilizes plots of *species* importance against environment, whereas *community* measures are used in the definitions given in Figure 1.

Some difficulty could be experienced in using direct gradient analysis for testing the plant community hypotheses. This is illustrated in Figure 2 where species importance is plotted along ranges of environment. Three possibilities are shown. Case (a) illustrates uniform distributions with overlapping species ranges. The interpretation of this case might favor the continuum hypothesis although such a uniform group of species curves has been found only rarely. In case (b), the species curves are plotted to illustrate three groups of clustered species with correspondence of species distributions and regions along the environment where several species terminate. From this case, "types" and "ecotones" could be postulated. The two "ecotones" may also be characterized by species with narrow ranges occurring in the regions where the major species terminate. Hypothetical case (c) is more like the typical published results (*c.f.*, WHITTAKER, 1956, 1967). A variety of expressions may occur. Species may have wide or narrow ranges of tolerance and high or low values of importance. In such a case, the interpretation may be difficult for testing the plant community hypotheses.

WHITTAKER (1956, 1967) and BEALS (1969) discuss the appearance of flattened or truncated species distributions similar to the species on the far right of case (c) in Figure 2. WHITTAKER (1967) attributes these "plateau distributions" to species with a strong competitive advantage over some subrange of environment and never finds them in populations of mixed dominance. BEALS (1969) found that truncated distributions on a steep but not on a gentle slope and interprets the plateau distributions as caused by vegetative reproduction and competitive exclusion. BEALS' interpretation is that vegetation may form both a continuum and discrete groupings depending upon the conditions (in this case steep or gentle slopes). Both WHITTAKER & BEALS utilized a range of elevation instead of an independent measure of an environmental complex. This is contrary to the conditions specified earlier in this section. Environment may not vary uniformly with elevation. Species curves tend to show species individuality but not community individuality. Hence, this evidence cannot be accepted as a final test of the plant community hypotheses.

Environment can be treated at three levels of integration. Firstly, environment acts at the organism level. Additionally, the individual plants have genetic properties which allow them to compete over ranges of many environmental variables so that environment must be considered at the species or perhaps ecotype

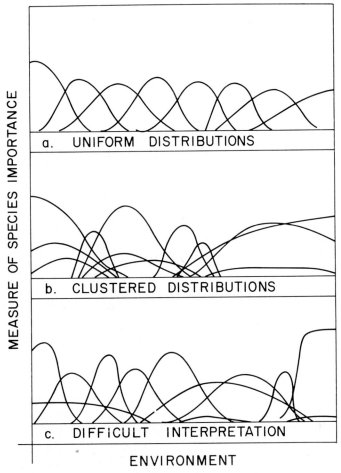

Figure 2. Hypothetical results of plotting species importance against a measure of community environment. See text for discussion of the three cases.

level. Finally, the "integrated" environment of groups of individuals of many species must be considered. The problem under discussion in this section requires definition of environment in this latter most highly integrated sense.

The concept of environment at the individual level is treated by MASON & LANGENHEIM (1957). Their definition emphasizes that the environmental relation takes place only at the organism level and lasts only while the individual organism lives. While this very restricted use is logical, it makes the measurement of environment a difficult procedure. The work of GATES (1965), MOEN (1968), and PORTER & GATES (1969) illustrates some of the attempts at estimating the environment of organisms.

The inclusion of only those variables which directly impinge on organisms at the organism level does not really rule out a determination of environment at the community level, but it points out that care must be exercised. Environment *acts* at the individual level but certainly higher levels of organization can be considered as the integration of all of the direct individual-environment relations.

The concept of environment in its most integrated (or holocoenotic) sense has been considered in many textbooks but in a rather loose manner. MAJOR (1951) and BILLINGS (1952) have attempted more rigorous treatments. An updated version of the statement by BILLINGS and its relationships with MASON & LANGENHEIM's (1957) "operational environment" appears earlier in this volume. Also D. SCOTT in another part of this book provides a more detailed and factorial view of the concept of environment. While the holocoenotic concept of environment may not properly be measured for some time, these attempts will serve as necessary foundations upon which future understanding must be constructed. Let us consider some of the problems of treating environment at the community level.

6.2.3   ENVIRONMENTAL FACTORS

MAJOR (1951) and BILLINGS (1952) have expressed the rudiments of the vegetation-environment relation in a manner similar to that in Figure 1. Vegetation is defined as a function of several "factors" of environment. Thus, we might consider the function

$$V = f(c, p, r, o, t) \tag{1}$$

where V is vegetation, and the letters within the function represent the factors of climate, parent material, relief, other organisms and time respectively. Treating the "factors" as independent variables, the total derivative can be obtained as a group of partial derivatives of the independent variables. Thus, one can consider vegetation in terms of one factor with all others held constant. For example, the relation of vegetation to the "climatic factor" would be

$$V = f(c)_{p, r, o, t} \tag{2}$$

where all of the other independent variables are held constant and only the "climatic factor" is considered. This is in fact what is implied in Figure 1 and what has been attempted in many studies. For example, WHITTAKER (1956, 1960) used moisture and elevation factors, LOUCKS (1962) used three factors of moisture, soil nutrients and local climate and MONK (1965) used factors derived from soil measurements.

While it may be unrealistic to consider the "factors" as independent there are statistical procedures which can be employed to treat them as independent. The procedure of obtaining results such as those in Figure 1 is in any case stochastic and not analytical since the lines are either means or fitted curves providing only analytical analogies.

Environmental factors in the sense of equations (1) and (2) may be derived from an exhaustive study of a large number of variables where the measurements relate to individual organisms. To preserve the concept that the environmental relation acts at the organism level, the methods such as used by GATES (1965) might be utilized and some procedure of "integration" found to derive the "community environment". Such a study will yield several major "factors" of the "environmental complex" in the sense discussed by BILLINGS (1952 and in this chapter). No matter how intensive the study or how reliable the "integration" procedure, no study of community environment can provide more than this. Therefore, profitable research for the study of community environment should emphasize the definition and measurement of the various important factors of the environmental complex. The vegetation should be studied over the largest ranges of these factors in order to test the plant community hypotheses.

6.2.4   SUCCESSION, SITE POTENTIAL AND ENVIRONMENTAL FACTORS

Even if the ambiguity of the concept of environment between the community and organism levels can be avoided, an additional difficulty arises when the environmental factors are considered over time. An example of recent disturbance and changing environment could be the effects of logging over a large part of a region. Suppose there are two nearby sites with differing characteristics of soil material and local climate. Allow that after the disturbance both sites receive essentially the same kinds and numbers of seed propagules so that in the very early stages of secondary succession, the vegetation on both sites is nearly the same. Succession may then proceed as in Figure 3 where the vegetation index represents one axis of an ordination. Because of the difference in potential of the sites, differing vegetation may result. Finally, when further changes are very small, stand (a) has changed more than stand (b). The example might represent a case where site (a) is mesic while site

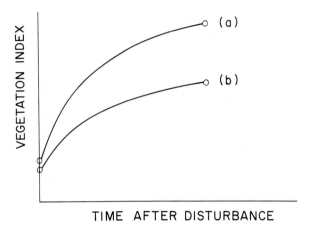

Figure 3. Hypothetical plots of the change in community vegetation index of two stands due to succession following disturbance.

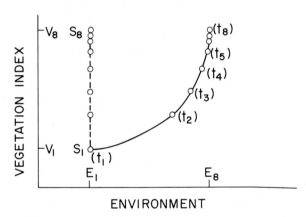

Figure 4. Hypothetical plot of the change in a community vegetation index (V), a community measure of environment (E) and the potential of a site (S) for a stand of vegetation due to succession following disturbance.

(b) is either somewhat more wet or dry. Other possibilities exist. The importance of the illustration is that the *potential* of a site is strongly related to the site environment. Can these be confused? If so, will this cause difficulties in hypothesis testing?

Part of the answer to these questions is illustrated in Figure 4. This shows the hypothetical change of vegetation on a site over time with respect to an index of an environmental factor. As the vegetation develops over equal intervals of time (t_1 through t_8) from V_1 to V_8 the index of the environmental factor measured for the stand would also be expected to change from E_1 to E_8. After a suitable time is reached (t_8) the vegetation will probably change very little with additional time. However, an index of the site *potential* would not be expected to change so that S_1 through S_8 are hypothetically identical. In practice it may be very difficult to distinguish site potential from the measure of the environmental factor used because the environment changes with succession. Perhaps an ideal index of environment for hypothesis testing would be some measure of site potential because it does not change with succession and would not differ as a result of chance factors. The measure of the environmental factor which can be realistically obtained in such a study would probably lie somewhere between site potential and environment on the environment axis of Figure 4. Research on defining site potential and its difference from the stand environment after some degree of steady state would be fruitful for testing of the plant community hypothesis.

6.2.5 SPACE-RELATED FACTORS

It was pointed out earlier that the relation between vegetation and space is not a question to be answered in the testing of the plant community hypotheses because environment does not necessarily change uniformly in space. Therefore, space-related factors such as topography should be considered with caution. In fact, BILLINGS (1952) points out that "topography" never affects an organism directly but only through other environmental components. The term "relief" in equations (1) and (2) should be suspect. In particular, ranges of elevation and slope direction have been used in studies to demonstrate the correctness of both plant community hypotheses.

An example will be used to demonstrate the difficulty of using elevation as an environmental factor. SCOTT & HOLWAY (1969) reported on a study of 180 stands of forest vegetation on Whiteface Mountain in the Adirondacks of New York. The stands were located over a range of elevations, slope aspects and slope steepness. Details of the sampling methods are contained in the original report. A majority of the stands were undisturbed by logging or fire for nearly 100 years and some for longer periods. This could not be said for the entire sample although we believe that none of the stands are less than 70 years old.

The data were used in several kinds of ordinations with the results being similar although not exactly the same for all methods. In Figure 5, a vegetation index is plotted against elevation of the

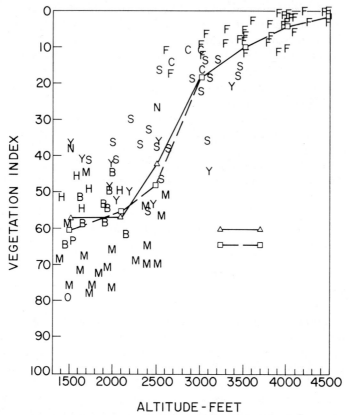

Figure 5. Scatter diagram of a vegetation index against altitude on Whiteface Mountain in the Adirondacks, New York. Lines connect mean values of the index for 500 foot intervals of altitude. Letters designate leading dominant of the stand. Acer saccharum-M (sugar maple), Fagus grandifolia-B (American beech), Tsuga canadensis-H (hemlock), Betula allaghaniensis-Y (yellow birch), Quercus rubra var. borealis-O (northern red oak), Populus grandidentata-P (large-toothed apsen), Thuja occidentalis-N (northern white cedar), Picea rubens-S (red spruce), Abies balsamea-F (balsam fir), Betula papyrifera var. cordifolia-C (chordate-leaved birch).

stand. The index in this case was obtained by essentially the same techniques as the "strip method" described by CURTIS & MC-INTOSH (1951).

In the scatter diagrams of Figure 5, the symbols refer to the first leading dominant of the stand. In this particular plot all pine-dominated and level stands have been removed because of the specialized site characteristics they represent. The mean values of the vegetation index for 500 foot ranges of altitude have been plotted for two different methods of averaging. In the plot shown by the solid line, each stand is given equal weight no matter how many stands occurred in a given range of slope-aspect. In the other plot (dashed line) the means are formed by averaging over quadrants of slope-aspect (N, E, S, W) which are then weighted equally to form the mean for each altitude range. Also, for the latter plot the stands outside of the immediate Whiteface Mountain area were not included whereas they were included in the first plot. There were several other methods of averaging which were tested but none of the results were very different than those shown in Figure 5. The curves were also fitted by regression analysis with the result being essentially the same as Figure 5. Using the same data, BREISCH, *et al.* (1969) also obtained essentially the same result utilizing the similarity index method of BRAY & CURTIS (1957) to obtain the vegetation index. These same data have since been analyzed by more rigorous techniques and regression analysis by ALAN AUCLAIR (unpublished) of McGill University. In all cases where the first axis of an ordination is plotted against altitude the results are essentially the same as the plots in Figure 5.

The interpretation that one might first make from Figure 5 is the same as for case (c) of Figure 1. That is, the data indicate some tendency for "association" within the spruce-fir communities at high altitudes and within the northern hardwoods at lower altitudes with an "ecotone" between the two. This interpretation is actually not much different from that of CURTIS (1959) and McINTOSH (1967) where the continuum was believed to be demonstrated *within* gross physiognomic types but not necessarily *between* them. Thus, this interpretation favors both of the plant community hypotheses.

It would be helpful to discover what causes the sharp change in vegetation on Whiteface Mountain at the 2500—3000 foot altitude range because the answer would provide a test of the plant community hypotheses. If interspecfiic competition is the cause of the "ecotone" then we must conclude that "association" is a real phenomenon at least when different physiognomic types are involved. Perhaps the spruce-fir communities restrict development of hardwood seedlings on the forest floor because of shading or some influence on the soil. They may also provide some positive influence on spruce and fir seedlings. The hardwoods might similarly modify the seedling environment to enhance hardwood development but not that of spruce and fir. Certainly, other possibilities exist.

On the other hand, the "ecotone" on Whiteface Mountain may be caused by non-uniform changes in environment over the elevational range in question. This would add evidence in favor of the continuum hypotheses. The main reason for using this particular example is to illustrate that care must be exercised when using space-related variables instead of independent measures of environment. For example, there might be a sharp change in the characteristics of the substrate in the interval from 2500 to 3000 feet. As elevation increases in this range, the slope steepness increases rather abruptly and the substrate changes from a fairly deep glacial parent material to one which is more shallow and rocky in character. Many climatic elements might similarly change rapidly at the "ecotone". The "cloud cap" of Whiteface frequently extends down to 3000 feet but infrequently below. The long term effect of climatic variation of several key elements could be the cause of the sharp change in vegetation.

SICCAMA (1968) found a similar sharp change of vegetation on Camels Hump in the Green Mountains. His extensive analysis and discussion of the change of environment with elevation led to conclusions similar to ours (SCOTT & HOLWAY, 1969). In particular, SICCAMA found that frost and cloud conditions might change abruptly in the "ecotone" but the soils did not.

It will not be easy to prove whether environment or vegetation dynamics cause sharp transitions or "ecotones". Before the answer is obtained we must also find an answer to the question: Is community environment quantifiable and can it be linearized?

6.2.6   ENVIRONMENTAL SYNTHESIS

For the purpose of testing the plant community hypotheses, the concept of environment has been relaxed to the point where it is probably measurable. First, we have allowed that a community level (or holistic) determination will be suitable for hypothesis testing even though environment acts at the organism level. After advising caution concerning the confusion between site potential and community level environment and about the change of environment with succession, we then advised once more against the use of space-related variables such as elevation along a mountain slope. The meaning of the "holistic nature" of environment is equivalent to a factor complex approach as discussed by BILLINGS (1952, 1973). How can such a complex be synthesized from actual measurements?

First, what measurements should be included in a synthesis of a factor complex? Actually the plants themselves provide the most convenient answer to this because natural plant communities are the best indicators of the physical environment. We could, therefore, use some simple form of vegetation analysis to pre-select the appropriate physical measurements. This is, of course, the procedure used in most studies seeking causality of plant community distribution. For example, a wet to dry sequence along some moisture factor complex may be indicated by the range of species in a study area. We would, therefore, want to measure edaphic and climatic variables related to how wet or dry the site may be. Many lists and explanations of the appropriate measurements are available including BILLINGS (1952), PLATT & GRIFFITHS (1964) and GEIGER (1965).

Some investigators (c.f., LOUCKS, 1962, and WARING & MAJOR, 1964) actually use evidence of a significant response from physiological studies before selecting appropriate measurements. Are these procedures circular? My opinion is that there is little danger of a circular argument when a factor complex is to be obtained provided enough different kinds of measurements are included in the synthesis.

The proof that circularity is not a question is difficult. It relies on the meaning we have chosen for the term "environment". Also, we have accepted a *community* level not a species or individual level. We want to discover how an index of community changes over a range of some synthesized factor complex of the environment. Choosing appropriate measures based on the fact that the component species show varied response with respect to the physical measurements will not endanger our calculation of the community environmental complex provided we test all of the species. Of course, circularity can be completely avoided by measuring all of the possible physical variables for synthesizing the environmental factor. Such a procedure would be unrealistically expensive.

Once a large number of appropriate environmental variables have been measured, the next step is to synthesize them into a linear form of one or several factor complexes. There is no certainty that the data can be linearized. LOUCKS (1962) has attempted to resolve this problem by scalar analysis where linearization is accomplished by selecting the most appropriate mathematical relation between the vegetation and environment. This procedure, while somewhat arbitrary, will ensure that unrealistic variables are not included in the synthesis. Regression analysis might be considered for linearizing the environmental measurements to avoid the use of somewhat arbitrary decisions.

Although research toward obtaining the factor complex in linearized form will be fruitful, the procedure of obtaining an environmental index might itself lead to linearization. This is because the environmental measurements tend to be correlated with each other. If a large number of variables related to one factor complex are included in a regression analysis against a measure of vegetation only a few will be required to obtain most of the variance which can be explained. A suitable environmental index might, therefore, be obtained from an ordination of the environmental measurements by one of the more suitable techniques such as factor analysis, principal components analysis or the similarity index method.

It is beyond the scope of this essay and the knowledge of the author to prove that a linear form of an environmental factor complex can be reliably obtained from a number of physical measurements. This discussion was included to emphasize the character of the measurements required and the means by which they must be related to the vegetation. That is, the environmental measurements must be independent of the vegetation and the relation should take the form of Figure 1 for hypothesis testing. Figure 1 can, of course, be represented for more than one variable although only one is required for each environmental factor complex to be tested.

6.3 The Contest

This essay grew out of an apprehension on the part of the author that the debate between the advocates of the two mutually exclusive "plant community hypotheses' was not understood by all ecologists. Before reading the comprehensive review by MCINTOSH (1967) and the responses to his paper (DANSEREAU, *et al.*, 1968) by ten other ecologists, I was unaware that the conditions underlying the "controversy" were not universally accepted. Several readings of this extremely provocative two-part experiment (which this author likes to call "DANSEREAU's concert" because of the analogy DANSEREAU used in this forward) led to a feeling that there were a great many unnecessary complications in this debate. Some of these were semantic, some were due to inconsistent or even improper use of simple mathematical terminology but others were based upon a basic disagreement on how to approach the problem. In my opinion, a simplified statement of the opposing hypotheses free from the vagaries of data interpretation was badly needed.

One of the recurrent themes of the recent literature on this debate has been that a suitable "test" of the plant community hypotheses should be conducted under mutually agreed upon conditions. It is, therefore, proper to end this section of a chapter on vegetation and environment with a proposal to hold a "contest" (a term suggested by GIMINGHAM, 1968).

The conditions which have been discussed for testing the plant community hypotheses have never been met in any study published so far. It is my opinion that the most comprehensive study so far published was by LOUCKS (1962) in New Brunswick. The weakness of LOUCKS' study was that it was not large enough to include measurement of all of the necessary environmental variables. The range of vegetation and environment was also somewhat restricted. LOUCKS (1962) concluded that the continuum hypothesis was most appropriate in his study area, but he did not specifically look for the subtle kind of variation which might indicate case (b) of Figure 1. A summary of the data in an analytic form or a more complete set of environmental measurements might have provided enough sensitivity to either detect or rule out tendency for "association".

It is proposed that a large-scale multi-disciplinary study of the relation between vegetation and environment be undertaken in suitable study areas under conditions mutually agreed upon by components of both the community-type and continuum hypotheses. Two kinds of study areas should be utilized. One should involve a large variation of environment over a small distance where more than one gross physiognomic type is represented. Obviously a mountain slope would be ideal for this. Some good examples are the San Francisco Peak area of northern Arizona (DAUBENMIRE, 1943), the Bitterroot Mountains of Washington (DAUBENMIRE, 1966), the Front Range of Colorado (MARR, 1961, 1968), the Siskiyou Mountains in Oregon and California (WHITTAKER, 1960) or the Santa Catalina Mountains of Arizona (WHITTAKER & NIERING, 1964 and HAASE, 1970).

Another kind of study area should be examined where a range of vegetation and environment is studied within one physiognomic type. Two good examples are the upland forests of Wisconsin (CURTIS, 1959) and the steppe of eastern Washington (DAUBENMIRE, 1966).

The critical part of the proposed study is not where it is done but the detail which is applied. The vegetation should be sampled by all accepted techniques by several groups of specialists. The environmental measurements should be obtained by a wide range of experts including meteorologists, agronomists, geologists, and environmentally oriented ecologists. Data treatment should be accomplished by all available techniques but should include synthesis of vegetation and environmental indeces for presentation in the form of Figure 1.

6.4 Perspective

This essay has been limited to a narrow part of vegetation science. However, much of what has been said applies to other aspects of vegetation analysis. Many ecologists prefer to study particular attributes of plant communities related to successional diversity, stability, competition, energetics, nutrient cycling and other aspects of population dynamics or species evolution (ODUM, 1969). An intensive study of the vegetation and environment of an area including a large number of samples over a large range of conditions provides a framework in which these community attributes can be studied with greater success. It is my opinion that environment should also be studied in the broader context of community environment.

One of the recurrent themes of "DANSEREAU's Concert" (MCINTOSH, 1967 and DANSEREAU, et al., 1968) concerned the vegetation- environment relation. However, throughout the review, the responses and the reply by MCINTOSH, this theme remained in a brooding minor key and was often hidden in the counterpoint of recent developments. It never became a fully-orchestrated dominant theme. The present essay has attempted to rectify this remission by placing environment in its proper role and in a major key.

Environment as a concept is extremely difficult to define. It may be nearly impossible to measure correctly. Perhaps this concept is the "aberrant daughter" of the family of ecologists who deal with vegetation. She may have gone astray and have a shrew-like disposition but no matter what her inconsistencies, we must not keep her hidden in the backroom of our science. We might discover her "secret torment" by allowing her better access to the front porch of free inquiry. Its makes no sense to speak of "noda', "recurrent arrays", "associations", "discontinua", or "continua" unless you first ask: With respect to what? Vegetation might be expressed as discontinuous in space or even with respect to time but still represent a continuum with respect to *environment*. The coordinate system must be specified in an unambiguous manner.

MCINTOSH (1968) and CANTLON (1968) have specified some conditions which must be agreed upon for testing the plant community hypotheses. They did not specify that a measure of community environment be included in the test because they did not utilize environment in any statement of the hypotheses. Of the ten respondents in DANSEREAU, et al., (1968) only LIETH (1968) emphasized that the "key-lock" relationship requires an understanding of environment although he connects environment to communities as a controlling factor of species distribution similar to the gradient analysis technique of WHITTAKER (1967). In his reply to LIETH, McINTOSH (1968) states: "With respect to gradient analysis, my earlier suggestion (McINTOSH, 1958) was that it refer to ordinations based upon data of the physical environment whereas ordinations based on the vegetational composition be called continuum analysis. I believe now that the term ordination should cover both. The end, of course, is to seek coincidence between environmental and compositional ordinations". This statement suggests what this present essay has emphasized. The proper test for the plant community hypotheses must discover the function relating vegetation and environment both of which must be expressed in their holocoenotic (community) form by means of data processing such as ordination.

It was emphasized in our discussion of community environment that there is a danger of circular reasoning implied in the definition of environment. This is partly overcome by avoiding confusion between measurement of environment at the individual, species and community levels. Once we agree that the community level is appropriate we must also distinguish between measurement of site and of the changed environment of a site as illustrated in Figures 3 and 4. How does environmental conditioning of a site differ from "potential environment?" Which measure do we use and how will the choice influence the testing of our stated hypotheses? The answers to these difficult questions will provide further insight into the concept of the "web of life".

REFERENCES

AUSTIN, M. P. & L. ORLOCI – 1966 – Geometric models in ecology. II. An evaluation of some ordination techniques. *Jour. Ecol.* 54 (1): 217–227.
AYYAD, M. A. G. & R. L. DIX – 1964 – An analysis of a vegetation-microenvironmental complex on prairie slopes in Saskatchewan. *Ecol. Monogr.* 34: 421–442.
BEALS, E. W. – 1969 – Vegetational change along altitudinal gradients. *Science* 165: 981–985.
BILLINGS, W. D. – 1952 – The environmental complex in relation to plant growth and distribution. *Quart. Rev. Biol.* 27: 251–265.
BILLINGS, W. D. – 1973 – Environment: Concept and reality. Handbook of Vegetation Science (this volume).
BRAY, J. R. & J. T. CURTIS – 1957 – An ordination of the upland forest communities of southern Wisconsin. *Ecol. Monog.* 27: 325–349.
BREISCH, A. R., J. T. SCOTT, R. A. PARK & P. C. LEMON – 1969 – Multi-dimension ordination of boreal and hardwood forests on Whiteface Mountain. Report No. 92. Atmospheric Science Research Center, SUNY at Albany, Albany, New York.

CANTLON, J. E. – 1968 – The continuum of vegetation: Responses. *Bot. Rev.* 34: 255–258.
CURTIS, J. T. – 1959 – The vegetation of Wisconsin: An ordination of plant communities. 657 pp. Univ. of Wisconsin Press, Madison, Wisconsin.
CURTIS, J. T. & R. P. MCINTOSH – 1951 – An upland forest continuum in the prairie-forest border region of Wisconsin. *Ecology* 32: 476–496.
DANSEREAU, P. (with ten authors) – 1968 – The continuum concept of vegetation: Responses. *Bot. Rev.* 34: 253–332.
DAUBENMIRE, R. F. – 1943 – Vegetational zonation in the Rocky Mountains. *Bot. Rev.* 6: 325–393.
DAUBENMIRE, R. F. – 1966 – Vegetation: identification of typal communities. *Science* 151: 291–298.
GATES, D. M. – 1965 – Energy, plants and ecology. *Ecology* 46: 1–13.
GEIGER, R. – 1965 – The climate near the ground. 611 pp. Harvard Univ. Press. Cambridge, Massachusetts.
GLEASON, H. A. – 1939 – The individualistic concept of the plant association. *Amer. Midl. Nat.* 21: 92–110.
GIMINGHAM, C. H. – 1968 – The continuum concept of vegetation: Responses. *Bot. Rev.* 34: 273–290.
GOFF, F. G. & G. COTTAM – 1967 – Gradient analysis: the use of species and synthetic indeces. *Ecology* 48: 793–806.
GREIG-SMITH, P. – 1964 – Quantitative plant ecology. 2nd Ed. Butterworths and Co., London, xii+256 pp.
HAASE, E. F. – 1970 – Environmental fluctuations on south facing slopes in the Santa Catalina Mountains in Arizona. *Ecology* 51: 959–974.
LAMBERT, J. M. & M. B. DALE – 1964 – The use of statistics in phytosociology. In: "Advances in Ecological Research", Academic Press, Inc., London, Vol. II, pp. 55–99.
LANGFORD, A. N. & M. F. BUELL – 1969 – Integration, identity and stability in the plant association. *Adv. in Ecol. Res.* 6: 83–135.
LIETH, H. – 1968 – The continuum concept of vegetation: Responses. *Bot. Rev.* 34: 291–302.
LOUCKS, O. L. – 1962 – Ordinating forest communities by means of environmental scalars and phytosociological indices. *Ecol. Monogr.* 32: 137–166.
MAJOR, J. – 1951 – A functional factoral approach to plant ecology. *Ecology* 32: 392–412.
MARR, J. W. – 1961 – Ecosystems of the east slope of the front range in Colorado. Univ. Colorado Studies, Series in Biol. No. 8.
MARR, J. W. – 1968 – Data on mountain environments. II. Front Range Colo., four climax regions, 1953–1958. Univ. Colo Studies. Series in Biol. No. 28.
MASON, H. L. & J. H. LANGENHEIM – 1957 – Language analysis and the concept of environment. *Ecology* 38: 325–339.
MAYCOCK, P. F. & J. T. CURTIS – 1960 – The phytosociology of boreal conifer-hardwood forests of the Great Lakes region. *Ecol. Monogr.* 30: 1–35.
MCINTOSH, R. P. – 1958 – Plant communities. *Sci.* 128: 115–120.
MCINTOSH, R. P. – 1963 – Ecosystems, evolution and relational patterns of living organisms. *Amer. Sci.* 51: 246–267.
MCINTOSH, R. P. – 1967 – The continuum concept of vegetation. *Bot. Rev.* 33: 130–187.
MCINTOSH, R. P. – 1968 – The continuum concept of vegetation: Responses (reply to comments). *Bot. Rev.* 34: 315–332.
MOEN, A. M. – 1968 – Energy exchange of white-tailed deer, eastern Minnesota. *Ecology* 49: 676–681.

Monk, C. D. – 1965 – Southern mixed hardwood forest of North Central Florida. *Ecol. Monogr.* 35: 335–354.
Odum, E. – 1969 – The strategy of ecosystem development. *Science* 164: 262–270.
Platt, R. B. & J. F. Griffiths – 1964 – Environmental measurement and interpretation. 235 pp. Reinhold Publ. Corp., New York.
Porter, W. P. & D. M. Gates – 1969 – Thermodynamic equilibria of animals with environment. *Ecol. Monogr.* 39: 225–244.
Siccama, T. G. – 1968 – Forest zonation, soils and climate in the Green Mountains of Vermont. Ph. D. Thesis, Univ. of Vermont, Dept. of Botany.
Scott, J. T. & G. G. Holway – 1969 – Comparison of topographic and vegetation gradients in forests on Whiteface Mountain. Report No. 92. Atmospheric sciences Research Center, SUNY at Albany, Albany, New York.
Waring, R. H. & J. Major – 1964 – Some vegetation of the California coast redwood region in relation to gradients of moisture, nutrients, light and temperature. *Ecol. Monogr.* 34: 167–215.
Whittaker, R. H. – 1956 – Vegetation of the Great Smoky Mountains. *Ecol. Monogr.* 26: 1–80.
Whittaker, R. H. – 1960 – Vegetation of the Siskiyou Mountains, Oregon and California. *Ecol. Monogr.* 30: 279–338.
Whittaker, R. H. – 1967 – Gradient analysis of vegetation. *Biol. Rev.* 49: 207–267.
Whittaker, R. H. & W. A. Niering – 1964 – Vegetation of the Santa Catalina Mountains, Arizona. I. Ecological classification and distribution of species. *Jour. Ariz. Acad. Sci.* 3: 9–34.

7 PLANT FORMS IN RELATION TO ENVIRONMENT

H. A. Mooney

Contents

7.1	Introduction	113
7.2	Form and Function	114
7.3	Environmental Gradients and Plant Forms	116
7.4	Niche Specialization	117
7.5	A Test of the System	119

7 PLANT FORMS IN RELATION TO ENVIRONMENT

7.1 Introduction

In any given climatic region, the natural landscape is covered by plant communities which are usually dominated by a single growth form [1]. As examples one can cite the needleleaf evergreen trees of the cold winter climates, broadleaf evergreen trees of warm moist climates, and evergreen sclerophyll shrubs of the winter rain-summer drought climates. However, there may be a mosaic of physiognomic types within a particular climatic area related to the overriding influence of certain environmental variables such as substrate diversity (MAJOR, 1967). For example, grasslands may occupy soils derived from serpentine rock adjacent to woodlands on sandstone soils, all within the same climatic realm. Further, within a uniform macroclimate and substrate type there may be dissimilar physiognomic units present which represent a successional series, such as is often the case, from an herbaceous annual, to herbaceous perennial to a woody shrub, and finally forest type.

Examination of any one of these communities shows that the dominants usually belong to completely different genera or even families. These "look-alikes" are thus not necessarily closely related. The familiar winter deciduous trees, the oaks, maples, and beeches of the northern temperate forest may be cited as a single example. The multiplicity of species with the graminoid growth form in the world's grasslands is another.

Not only do many of the dominants of a given community often look alike, but, as has been known for a long time, where climates repeat, similar forms will be found. Thus, areas which are climatically similar, but geographically isolated, will be characterized by similar plant forms which in total have had very different evolutionary histories. The classic example of such a convergence is the evergreen sclerophyllous scrublands of the world's mediterranean climatic regions.

It thus appears that there is an optimal dominant growth form

---

[1] The term "growth form" is used here as morphological appearance of a plant as well as certain behavioral characteristics such as leaf duration. A community of plants dominated by a given growth form are of a single physiognomic type.

for a given climatic-substrate-successional combination. These observations are as old as the science of ecology and have served as the foundation of many "schools" within this science as well as for certain aspects of plant geography and paleoecology. Many classifications of climate are based on the correspondence of climatic elements with plant physiognomic types. Further, statements about paleoclimates have been made on the basis of the leaf sizes and shapes, in addition to other morphological features. As CAIN (1944) has stated, "even without botanical identification, the morphology of fossilized plants is a useful indication of environmental conditions". This of course applies to modern plants and environments.

The sum of all this is that to a large degree one need not know the details of the floristics of a region in order to make meaningful interpretations of the environmental relationships of the extant organisms. Further, since these forms and environments are repeatable, predications can be made.

The examples given so far have been directed toward the dominants of a community. It can be shown also that there often is a similar degree of convergence in the subordinate units of the vegetation. RAUNKIAER's (1934) long-used classification of the forms of plants based on the position of perennating buds (life forms) has demonstrated that when considering the entire flora of a given climate type there will be a similarity in life-form composition. This means, in a general sense, that the total kinds of adaptive strategies found in any given community are limited. According to BAKER (1966) this vegetational structural similarity in homologous climates extends to such functional features of the community as seed dispersal mechanisms and reproductive systems.

The above generalities are among the oldest observations of ecology, and have served as unifying concepts of the science. However, the details of the interrelationships of plant forms, function, and environment are as yet poorly understood.

7.2 Form and Function

LIVINGSTON & SHREVE (1921) in their treatise on the distribution of the vegetation of the United States established the philosophical foundation for modern plant physiological ecology. They stated that a classification of "the plants which have solved the same problems of environmental adjustment in the same manner" would be very different than a phylogenetic classification. They thought that the various growth-form systems which had been proposed up to that time, which were based on anatomy and physiology, were the beginnings of an ecological classification of plants from a physiological standpoint which would be perfected in the coming years. All that was needed was further comparative physiological work and the experimental demonstration of the adaptive significance of diverse morphological features. Unfortunately, progress toward attaining this goal has been slow. For one reason, the science of comparative plant physiology never developed as anticipated. Further, and perhaps more important, the study of the ecological significance of plant morphological features — or ecological anatomy — which flourished in the early 1900's suffered a severe setback with MAXIMOV's publications of 1929 and 1931. Briefly stated, post-Darwinian interpretations of the adaptive significance of the morphological features of organisms were often based mainly on speculation and infrequently on experimentation (BEWS, 1927). It was assumed, for example, that the anatomical features found in desert plants were those which restricted water loss. MAXIMOV, among others, however, showed that certain desert plants could have quite high transpiration rates[2].

It appears that these experimental findings more or less halted the study of the relationships between anatomical form and environment for the time and hence any progress toward an understanding of the basis of convergent evolution. Subsequent experimental studies on the adaptive significance of various plant structures have been very slow in developing.

Studies are slowly accumulating which are detailing the ecological significance of certain anatomical and structural features. As examples, one can cite the study of GATES (1968) on the relationship between leaf size, transpiration, and temperature and that of WUENSCHER (1970) on the influence of leaf hairs on plant temperature and water balance. These studies are providing a substantial framework of interpretation of form and function of plants.

Although it may be possible someday to interpret all plant forms in an adaptive-evolutionary context, it is important to realize that not all environmental adaptations of plants are manifested structurally. For example, differential biochemical adaptations to daylength, thermal regime, and soil nutrients may be found within a single morphological type as has been amply demonstrated in many genecological studies (HESLOP-HARRISON, 1964). However,

---

[2] More recent studies on the water loss characteristics of desert plant (CUNNINGHAM & STRAIN, 1969, MOONEY, et al., 1968) have shown that many species have exceptionally high rates of water loss which correspond to equally high rates of carbon gain. It is only when water is available in the habitat that these high gas exchange rates prevail. When water becomes limiting, the transpiring surfaces are either modified or lost.

most adaptive features of plants, particularly those related to gas exchange, involve structural as well as biochemical differentiation.

There is clear evidence that there are limited adaptive possibilities to any environment. Although it may not be possible in all cases to determine the habitat origin of a plant by its form alone it may certainly be possible by additionally considering its physiology or biochemistry.

COLE's (1967) comparative ecological study of the genus *Eriogonum* has important bearing on this problem. He found that members of closely-related species, which occupied the same habitat, were physiologically more similar than different races of these species in dissimilar habitats.

Similar parallel evolution of races of plants has been described by CLAUSEN (1962). In this case, morphological as well as physiological features were involved. Two different species, *Achillea borealis* and *A. lanulosa*, both have coastal bluff races which are short-stemmed and late flowering in comparison to other races of these species. TURESSON (1922) in his very early studies showed how coastal ecotypes have been repeatedly derived from woodland races of *Hieracium umbellatum*. All of these studies point to the limitations of adaptation to comparable selection, particularly within the same or similar genetic systems.

7.3 Environmental Gradients and Plant Forms

Although eventually it may be possible to predict the morphological-physiological combinations which are possible in any given environmental complex, we are still a long way from this sort of understanding. However, there is considerable information on the kinds of plant forms, and something of their function, which are found in different environments through space and time. The diverse forms found along gradual environmental gradients have provided material for an interpretation of the basis of form-environment relationships (AXELROD, 1966, WEBB, 1959, BEARD, 1944). MOONEY & DUNN(1970a) have recently interpreted the vegetational change found along aridity gradients in mediterranean climates primarily on the basis of carbon and water balance of the plants. In the regions analyzed, there is a progression along a gradient from the moistest to the driest climates of communities dominated by evergreen trees, evergreen shrubs, drought deciduous shrubs, and finally a mixed community of drought deciduous and succulent elements. These trends were noted along essentially identical climatic gradients in both California and Chile and hence they offer support for the concept of limited adaptive solutions to a given environmental complex. The argument was presented that there are energy costs involved in drought resistance which are repaid by potential carbon gain in those regions where moisture is only limited for a short period. Longer drought periods makes this investment non-competitive with drought evaders.

A detailed consideration of the trends found in the world's climatic types in terms of gas exchange capacity may provide a unifying concept of understanding. The balance between a morphological-behavioral system which has the maximum capacity for carbon fixation with a minimum water expenditure under the prevailing climatic conditions may be the prevalent adaptive mode for the region. Such an analysis may be complex indeed since it must consider all habitat features which influence gas exchange as well as the possible plant strategies for capturing and storing carbon and water.

It should be emphasized that although viewing the structural features of plants in terms of their potential for maximum carbon gain under the prevailing habitat conditions is of fundamental importance in understanding form-environmental relationships, a broader view of the selective forces of the habitat must also be maintained. For example, in an analysis of the selective forces which operate to produce the mediterranean-climate evergreen sclerophyllous shrub form it was necessary to consider, in addition to those features which are adaptive responses to maximize water and carbon capture and storage, those which are involved in mineral conservation and predator protection, as well as adaptive response to fire (Fig. 1) (MOONEY & DUNN, 1970b).

7.4 Niche Specialization

For any given habitat condition there is a plant type which apparently represents the optimal form-behavioral strategy for carbon gain. These types will be the dominant plants. In most environments, there apparently are only a few successful basic structural types although the number of genetic systems attaining this type may be very numerous in many tropical and temperate climates. There are certain habitats however where there are a multiplicity of quite diverse dominant plant strategies. SHREVE (1936) and WHITTAKER & NIERING (1965) have noted the increase in the number of growth forms found along increasing aridity gradients in the western United States. MOONEY & DUNN (1970a) have proposed that in those habitats where the vegetation is open

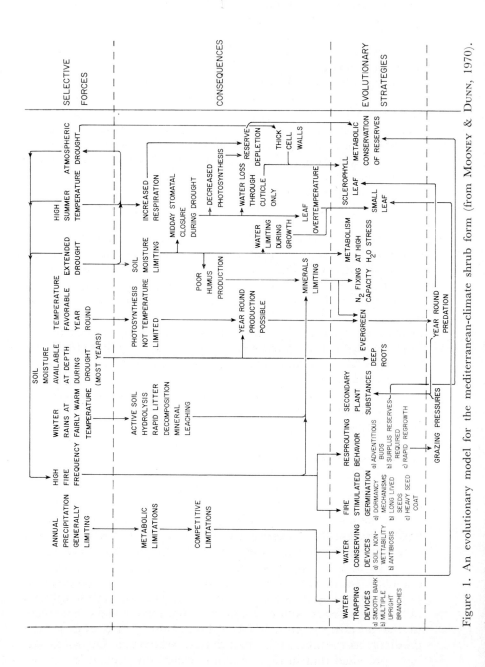

Figure 1. An evolutionary model for the mediterranean-climate shrub form (from Mooney & Dunn, 1970).

and light is not limiting, and thus where there need not be selection for rapid growth rate, there can be a greater variety of morphological types. Very slow growing, but highly conservative water users, such as succulents, can co-exist with mesophytic drought-evaders and highly drought tolerant evergreens.

The dominants of any vegetation only use a portion of the resources available. There does not appear to be any plant type which can capture all of the light, minerals, and water of its habitat. Such generalization of behavior would not be competitive with more specialized types adapted to efficiently utilize only a certain segment of the resources. For example, in mediterranean climatic areas, rainfall is limited primarily to winter and spring. Deep-rooted drought-enduring shrubs can utilize soil water resources at depth during the drought. The water resources at the very upper layers are quite temporal and are mostly utilized by an entirely different adaptive mode — the annual plant drought evaders. Resource division of this type is common in all communities. It is proposed here that the dominant structural plant type for any given habitat is the design which leads to the greatest carbon gain and which thus utilizes the most of the habitat resources. This design cannot competitively utilize all of the habitat resources however. The remaining resources constitute a habitat type which appears with high frequency in homoclimates. As an example, the closed broadleaf evergreen forests of certain types of temperate climates produce habitats where light energy is at a very low level. Specialized genetic systems can capture this limited resource. They have evolved repeatedly from different genetic stock wherever this environment occurs. Thus, one can expect convergencies in the form-behavioral systems in the subordinate species of a community wherever basic climatic types are repeated.

7.5 A Test of the System

Studies are now in progress which should considerably further our understanding of the basis of convergent evolution. The United States International Biological Program on the Origin and Structure of Ecosystems includes an analysis of the structural features of ecosystems which occupy desert and mediterranean-type homoclimates in North and South America. These studies are focusing on identifying the biological resources of the environments and determining how they are partitioned by the biota. The results of these studies should tell us, in detail, how dissimilar genetic stock has responded to similar selective forces. If these environments are homologous in all respects, and if there are only a finite number of adaptive possibilities to any combination of selective forces, predictions of the number and biomass of the various structural-functional adaptive types should be possible. For example, it can be reasoned that there will be a specific plant morphological-behavioral type which is dominant and which is most efficient at balancing carbon gain and water loss in these environments. This type, due to environmental limitations, will supply energy to

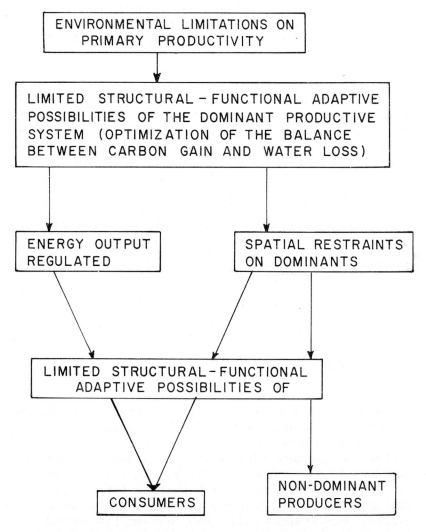

Figure 2. The influence of environment on structuring the ecosystem.

consumers in specific amounts and at specific times. Further, by its structural nature this kind of plant will provide a specific microclimate and shelter for other organisms. These in turn have an influence on the total kinds of adaptive types which may be possible in the environment (Fig. 2). Thus, it may be posible to identify not only the ecological homologs in the two continents, but to predict also the total numbers of any adaptive type.

ACKNOWLEDGMENTS

This study is an outgrowth of research conducted under NSF grants GB21441 and GB27151.

REFERENCES

AXELROD, D. – 1966 – Origin of deciduous and evergreen habits in temperate forest. *Evolution* 20: 1–15.
BAKER, H. G. – 1966 – Reasoning about adaptations in ecosystems. *Bioscience* 16: 35–37.
BEARD, J. S. – 1944 – Climax vegetation in tropical America. *Ecology* 25: 127–158.
BEWS, J. W. – 1927 – Studies on the ecological evolution of angiosperms. *New Phytol.* 28: 1–21, 65–84, 129–148, 209–248, 273–294.
CAIN, S. A. – 1944 – Foundations of Plant Geography. Harper and Bros. New York, 556 p.
CLAUSEN, J. – 1951 – Stages in the Evolution of Plant Species. Cornell Univ. Press. Ithaca, New York, 206 p.
COLE, N. – 1967 – Comparative physiological ecology of the genus *Eriogonum* in the Santa Monica Mountains in Southern California. *Ecol. Monog.* 37: 1–24.
CUNNINGHAM, G. & B. STRAIN – 1969 – Ecological significance of seasonal leaf variability in a desert shrub. *Ecology* 50: 400–408.
GATES, D. – 1968 – Transpiration and leaf temperature. *Ann. Rev. Plant Physiol.* 19: 211–238.
HESLOP-HARRISON, J. – 1964 – Forty years of genecology. In *Advances in Ecological Research* (ed.) J. B. CRAGG. Vol. 2: 159–247.
LIVINGSTON, B. E. & F. SHREVE – 1921 – The distribution of vegetation in the United States as related to climatic conditions. *Carnegie Institution of Wash. Publ.* 284, 585 p.
MAJOR, J. – 1967 – Potential evapotranspiration and plant distribution in western states with emphasis on California. In Ground Level Climatology. (ed.) R. H. SHAW. 93–126. AAAS.
MAXIMOV, N. A. – 1929 – The Plant in Relation to Water. Trans. by R. H. YAPP. George Allen and Univin Ltd. London, 451 p.
MAXIMOV, N. A. – 1931 – The physiological significance of the xeromorphic structure of plants. *Journ. Ecol.* 19: 273–282.
MOONEY, H. A., R. BRAYTON & M. WEST – 1968 – Transpiration insensity as related to vegetation zonation in the White Mountains of California. *Amer. Midl. Nat.* 80: 407–412.

Mooney, H. A. & E. L. Dunn – 1970a – Photosynthetic systems of mediterranean-climate shrubs and trees of California and Chile. *Amer. Nat.* 104: 447–453.

Mooney, H. A. & E. L. Dunn – 1970b – Convergent evolution in mediterranean-climate evergreen sclerophyll shrubs. *Evolution* 24: 292–303.

Raunkiaer, C. – 1934 – The Life Forms of Plants and Statistical Plant Geography. Oxford, 632 p.

Shreve, F. – 1936 – The transition from desert to chaparral in Baja California. *Madroño* 3: 257–264.

Turesson, G. – 1922 – The genotypical response of the plant species to the habitat. *Hereditas* 3: 211–350.

Webb, L. J. – 1959 – A physiognomic classification of Australian rain forest. *Jour. Ecol.* 47: 551–570.

Whittaker, R. H. & W. A. Niering – 1965 – Vegetation of the Santa Catalina Mountains, Arizona: A gradient analysis of the south slope. *Ecology* 46: 429–452.

Wuenscher, J. E. – 1970 – The effects of leaf hairs of *Verbascum thapsus* on leaf energy exchange. *New Phytologist* 69: 65–73.

8 MODELING THE PHOTOSYNTHESIS OF PLANT STANDS

C. E. Murphy, T. R. Sinclair & K. R. Knoerr

	Contents	page
8.1	Introduction	125
8.2	Leaf Process Models	127
8.2.1	Leaf Photosynthesis	127
8.2.2	Leaf Energy Balance	131
8.3	Stand Microclimate Models	133
8.3.1	Radiative Transfer	134
8.3.2	Convective Transfer	136
8.4	Stand Photosynthesis Model	138
8.5	Simulation Results and Discussion	140
8.6.1	Summary	142
8.7	Appendix I	144
8.7.1	Symbol List	144
8.7.2	Subscripts	145
8.7.3	Greek Letters	145

8 MODELING THE PHOTOSYNTHESIS OF PLANT[1,2,3] STANDS

8.1 Introduction

Photosynthesis is accomplished through a complicated sequence of enzyme-mediated reactions within chloroplasts. The quantity of carbon dioxide fixed in this process is determined by the rates of these various reactions. The reaction rates, in turn, are influenced by many factors, with the environments around the chloroplasts and leaves being the most important. One approach to investigating the relationships between photosynthesis and the environment, particularly that of whole plant stands, is through the use of mathematical models and computer simulation.

However, modeling of plant stand photosynthesis is not easy because of the many interactive processes which take place simultaneously. Meteorological conditions above the stand are the most important factors controlling the environment within the stand, yet information about these factors is generally not sufficient to predict photosynthesis. The plants themselves alter the stand environment by the many physical and physiological processes which take place at the plant-atmosphere and plant-soil interfaces. For these reasons we feel it necessary to incorporate many details of the processes which influence the environment of the plant stand and the photosynthesizing chloroplasts in the development of a model for stand carbon fixation.

Until recently it has been difficult to describe complex interactive systems such as stand photosynthesis. The difficulties were a result of both problems in finding mathematically tractable relationships to describe these systems, and the large amount of computation necessary to solve the relationships. Furthermore, and not least important, the underlying processes were not completely deciphered.

---

[1] Research supported by the Eastern Deciduous Forest Biome, US-IBP, funded by the National Science Foundation under Interagency Agreement AG-199, 40-193-69 with the Atomic Energy Commission-Oak Ridge National Laboratory.

[2] Contribution No. 101 from the Eastern Deciduous Forest Biome, US-IBP.

[3] A complete list of symbols used in this paper appears as Appendix I at the end of the paper.

However, advances in electronic computers and numerical approximation techniques during the last decade now make it possible to handle large amounts of computation in a reasonable amount of time to find solutions to many types of complex mathematical systems. While our detailed understanding of some of the processes is still not what we might desire, we feel that it is adequate enough for use in developing some preliminary models of complex systems such as stand photosynthesis.

The general procedures, for the use of modeling to integrate processes and determine the characteristics of the larger systems, have been developed thoroughly by Ashby (1956), von Bertalanffy (1968), Martin (1968), Patten (1971) and others. The basic approach outlined in our paper is that developed in detail by Martin (1968). Most of the steps in this approach are clearly identical with the more general precepts of "scientific methodology".

The first step in this approach is to answer the question, what is the objective of modeling this system? Often asking the right question is the most important part of the whole process of model construction. For this paper, we have defined our problem as determining the response of the photosynthesis of plant stands to change in the environment. The objective of constructing the model was to better understand the various interactive processes and their ultimate effect on carbon dioxide fixation. While it is probably premature to expect a stand photosynthesis model to have great utility as a predictive tool, we hope that this model will contribute to later models that can make predictions actually usable in the management of plant stands.

After the question of why the model is needed has been answered, the physical boundaries of the system being modeled must be defined. The natural, physical boundaries of the photosynthesizing stand are the upper limit of the plant canopy in the atmosphere and the lower limit of plant roots in the soil. For practical reasons, which will be given later, we have chosen to extend the upper boundary to a plane, in the upper part of the surface atmospheric boundary layer, some 10 meters above the stand, and restrict the lower boundary to the soil surface under the aerial part of the stand.

One of the most difficult steps is definition of the important subsystems of the total system in terms of usable mathematical expressions. The crucial subsystem of the stand is the leaf. This subsystem needs to be defined mathematically from two viewpoints. First, a reasonable expression for leaf photosynthesis in terms of various environmental components is needed. Second, those processes which determine the energy environment of the leaf must be considered. These processes are the absorption of radiant energy and its partitioning between reradiation and sensible and latent heat exchanges.

The microclimate of the whole plant stand is also a crucial subsystem for a photosynthesis model. Two groups of processes are important in producing the plant stand microclimate for photosynthesis. First, the divergence of the radiation flux densities must be calculated to predict the radiant energy available for leaf photosynthesis and leaf energy balance at different depths in the canopy. Secondly, the convective processes within and above the canopy must be evaluated to determine the vertical profiles of the meteorological parameters affecting the photosynthetic environment and energy balance at the leaf surface. The mathematical functions chosen to define each of these subsystems are presented later.

Taken together, these functions will give us the coupled set of equations necessary to model the plant stand photosynthesis system. Solution of this set of equations is difficult; sophisticated numerical techniques must be used to show how plant stand photosynthesis varies as a function of environmental and stand conditions. We will present the results of several simulations to demonstrate the potential capability of a stand photosynthesis model. However, until the model has been validated by experimental data, it is only a complex hypothesis of how portions of the plant environmental and biological systems operate.

8.2 Leaf Process Models

8.2.1 LEAF PHOTOSYNTHESIS

Several leaf photosynthesis models have been proposed in recent years (STEWART, 1970; WAGGONER, 1969; LOMMEN, et al., 1971 and HALL, 1971). The model used here, a further modification of previous models, was developed by one of the present authors for incorporation in stand growth models with a minimum input of photosynthetic data (see SINCLAIR, 1972, for complete details of this model).

The biochemical relationships for this model are built around a simple set of equations similar to those presented by RABINOWITCH (1951)

$$A_c + CO_2 \ldots ACO_2 \tag{1}$$

$$NADP + I \ldots NADPH \tag{2}$$

$$A_c CO_2 + NADPH \ldots P_g \tag{3}$$

where:

A = CO_2 acceptor
I = irradiance of photosynthetically active radiation (μeinsteins)
k_1 = reaction rate constants
P_g = gross photosynthate
NADP = nicotinamide adenine dinucleotide phosphate, an electron carrier in the photosynthesis process, the reduced form is referred to as NADPH

Equation (1) required that the carbon dioxide concentration within the chloroplasts be known. Resistances in the pathway for diffusion of CO_2 to the chloroplasts can be used to solve for the chloroplast CO_2 concentration as a function of the atmospheric CO_2 concentration (C_a). The resistance analogue diffusion equation used for this solution is,

$$P_n = \frac{s\,C_a - C_c}{s(r_c + r_{sc}) + r_{cw} + r_p} \tag{4}$$

where:

s = solubility of CO_2 in water
r_c = boundary layer resistance to CO_2 transfer
r_{sc} = stomatal diffusion resistance for CO_2
r_{cw} = cell wall resistance
r_p = protoplasm resistance
C_a = air CO_2 concentration
C_c = chloroplast CO_2 concentration
P_n = net photosynthesis

Combining the expressions for the biochemical reactions and the diffusional pathway, the following equation for gross photosynthesis was obtained.

$$P_g = \frac{-b - \sqrt{b^2 - 4ac}}{2a} \tag{5}$$

where:

$a = (k_2 I + 1)[k_1(r_p + r_{cw} + s(r_s + r_a))]$,
$b = (k_2 I + 1)[-sC_a k_1(r_{cw} + s(r_s + r_a)) - 1]$
    $- k_1 k_2 k_3 A_0 N_0 I(r_p + r_{cw} + s(r_s + r_a))$,
$c = k_1 k_2 k_3 A_0 N_0 I[R_s(r_{cw} + s(r_s + r_z)) + sC_a]$;

and:

R_s = mitochondrial respiration,
$A_0 = [A_c] + [ACO_2]$,
$N_0 = [NADP] + [NADPH]$.

Equation (5) can be solved if the values of the various parameters are known as constants or as some function of a measured environmental parameter or plant structural characteristics. A great deal is already known about some of the diffusion resistances. The boundary layer resistance has been shown to be defined by the relationship given in Equation (6) (GATES, 1968; PARKHURST, et al., 1968; THOM, 1968; STEWART & LEMON, 1969).

$$r_h = \text{constant}\ \left(\frac{D}{u}\right)^{\frac{1}{2}} \qquad (6)$$

where:

D = effective leaf width
u = wind speed
constant = 1.0—1.5 for heat transfer dependent on species and leaf angle.

To obtain diffusion resistances for CO_2 transfer, the values determined from Equation (6), which are for resistance to heat transfer must be multiplied by the ratio of the PRANDTL number to SCHMIDT number for CO_2 taken to the two-thirds power (0.94).

The stomatal resistance has also been calculated as a function of environmental parameters (GAASTRA, 1959; SHAWCROFT, 1970)

$$r_s = \frac{\beta}{I+I'} + \alpha \qquad (7)$$

where:

α = stress factor dependent on temperature and water

β, I' = constants which are a function of species and probably preconditioning.

A more comprehensive stomatal submodel than equation (7) would be very desirable. Definition of stomatal resistance in terms of interaction between various environmental parameters and the physiological processes that determine stomatal behavior should improve greatly the power of the entire plant stand model.

An estimate of cell wall resistance can be calculated from the equation presented by HALL (1971).

$$r_{cw} = \frac{tF}{d_w A_{cw}} \qquad (8)$$

where:

t = thickness of the cell wall (0.2—0.3 microns)
F = tortuosity factor (1—3.14)
d_w = diffusivity of CO_2 in liquid water (1.98×10^{-5} cm²/sec)
A_{cw} = projected area of the cell walls to area of leaf (10—30)

The final resistance, r_p, is difficult to calculate directly. The resistance to diffusion of CO_2 through the plasmalemma, protoplasm, and chloroplast membranes are lumped into r_p. Since a theoretical method for estimating r_p is not currently available, this resistance must be calculated using experimental data for the leaves being modeled.

The biochemical parameters, k_1, k_2, k_3, A_o, and N_o are also difficult to obtain by any direct method. However, readily available experimental data can be used to estimate these parameters for a range of environmental conditions. The light compensation point and light-saturated photosynthetic rate at a known temperature and CO_2 concentration provide the information necessary for their evaluation. However, our present ability to extrapolate the leaf photosynthesis model to various water and temperature stress conditions is limited by the lack of data required to evaluate the reaction rate constants. These "constants" will certainly vary as the stress conditions change. We already have preliminary data which indicates large changes in k_1 as a function of temperature.

The final information required to solve equation (5) is mitochondria respiration. Several expressions for this respiration are available. For simplicity, the following equation, presented by WAGGONER (1969), was chosen because it requires very little input data to generate a solution.

$$R_s = R_x \exp\left[9000 \ln (Q_{10})\left(\frac{1}{T_x} - \frac{1}{T}\right)\right] \qquad (9)$$

where:

R_x = base respiration rate
T_x = base temperature (°K)
Q_{10} = rate of change of respiration in a 10° C temperature change.

Eventually more sophisticated models of respiration will be used to replace the above equation. A model making respiration dependent on photosynthesis in addition to various environmental parameters would be very desirable. However, since the data required to use such a model are generally unavailable, equation (9) was used in our simulations.

From the foregoing discussion we see that the solution of equation (5) depends on the input of several environmental parameters. These parameters include atmospheric CO_2 concentration, light intensity, wind speed and air temperature. The numerical values for these inputs are obtained through the solution of the microclimate model. Light intensity is used as a direct input as well as in the evaluation of stomatal resistance. Wind speed is used to calculate leaf boundary layer resistance. Leaf temperature is used to estimate mitochondria respiration from equation (9) and will probably be needed in more sophisticated simulations to solve for the biochemical reaction rate and stomatal resistance. Leaf temperature is obtained through a solution of the leaf energy balance model.

8.2.2 Leaf Energy Balance

The leaf energy balance is the sum of all energy flux densities toward or away from the leaf and any energy storage within the leaf. It can be expressed by the following equation.

$$S_u+S_l+L_u+L_l+R_u+R_l+E_u+E_l+H_u+H_l+M_u+M_l+Q=0 \tag{10}$$

where:

S = short-wave radiation irradiance, primarily of solar origin.
L = long-wave radiation irradiance, primarily of terrestrial origin
R = long-wave radiation emittance
E = latent heat flux density from evaporation of water
H = sensible heat flux density
M = metabolic heat storage from net photosynthesis
Q = sensible heat storage
u = upper side of the leaf
l = lower side of the leaf

Usually sensible heat storage and the net metabolic storage terms are small compared to the other terms and thus have little effect on leaf temperature. Neglecting these smaller terms has two important consequences. First, we are able to uncouple the energy balance from the net photosynthesis equations. While net photosynthesis will be a function of leaf temperature, leaf temperature will no longer be a function of photosynthesis. Secondly, we can reduce the leaf energy balance to a steady-state problem by excluding the time dependent storage terms. Thus leaf temperature will only be a function of the present state of the leaf environment and not dependent on recent past conditions.

The steady-state leaf energy balance was first developed by Raschke (1956) and has since been further developed by a number of authors (Gates, 1968; Knoerr & Gay, 1965). The energy balance equation can be expanded to give,

$$\alpha_s(S_u+S_l)+\alpha_L(L_u+L_l)-2\sigma T_l^4+\frac{2\rho c_p}{r_h}(T_l-T_a)+\rho L_t$$
$$\left[\frac{1}{(r_w+r_{sw})^u}+\frac{1}{(r_w+r_{sw})^l}\right](q_l-q_a) = 0 \qquad (11)$$

where:

- S = short-wave radiation irradiance
- L = long-wave radiation irradiance
- α_s = absorptivity for short-wave radiation
- α_l = absorptivity for long-wave radiation
- σ = STEFAN BOLTZMANN constant
- u = upper side of the leaf
- l = lower side of the leaf
- T_l = leaf temperature (K)
- ρ = density of air
- c_p = specific heat of air
- r_h = boundary layer diffusion resistance to heat transfer
- T_a = air temperature
- L_t = latent heat of vaporization for water
- r_w = boundary layer diffusion resistance to water vapor transfer
- r_{sw} = stomatal diffusion resistance for water vapor
- q_l = leaf specific humidity (gm/gm)
- q_a = air specific humidity (gm/gm)

The parameters of equation (11) can be classed as environmental inputs, given constants, measurable physical characteristics of the leaf, or unknowns for which functional relationships must be found relating them to other parameters belonging to the first three classes. The photosynthesis model requires that this equation be solved for leaf temperature.

The values of intercepted radiation, air temperature and air specific humidity are the environmental inputs. The STEFAN-BOLTZMANN constant and air density and specific heat can be treated as given constants. The latent heat of vaporization is a linear function of leaf temperature,

$$L_t = 750.2-0.567\ T_l. \qquad (12)$$

The boundary layer resistance to heat transfer, as previously defined by equation (6) is a function of wind speed and average leaf width. The boundary layer resistance to water vapor flux density is the ratio of the PRANDTL number and SCHMIDT number for water vapor to the two-thirds power (1.24) times the resistance for heat. The stomatal resistance has also been defined previously as a function of photosynthetically active radiation and plant water and/or temperature stress. The structural characteristics of the leaves which are necessary inputs to the model are the average leaf width and the leaf absorptivity. Leaf area is also necessary if the total flux density for a single leaf is a desired output.

The remaining unknowns are leaf temperature and leaf specific humidity. The specific humidity is evaluated at the walls of the mesophyll cells inside the leaf. These walls are normally wet and for practical purposes the leaf specific humidity is the saturation specific humidity at the leaf temperature and atmospheric pressure. A number of equations are available for computing this relationship. The most widely used is the GOFF-GRATCH equation.

$$q_l = \frac{0.662}{P} \left[7.95357242 \times 10^{10} \exp \left\{ -18.1982839 \left(\frac{373.16}{T_l}\right) \right. \right.$$
$$+ 5.02808 \ln \left(\frac{373.16}{T_l}\right) - 70242.18 \exp \left(\frac{-26.1205253}{(373.16/T_l)}\right)$$
$$\left. \left. + 58.0691913 \exp \left[-8.03945282 \left(\frac{373.16}{T_l}\right) \right] \right\} \right].$$

This equation is rather long and a number of shorter approximations are available; however, since the computation can be done on a high speed computer the use of the above expression is not a serious problem (see MURRAY, 1967).

Using equation (11) through (13), the leaf energy balance can be solved to provide the leaf temperature required by the leaf photosynthesis model. However, the solutions and vertical integrations for both the leaf energy balance and photosynthesis models require considerable knowledge about the immediate leaf environment. This environment is defined by the vertical profiles of the irradiance of short-wave, long-wave and photosynthetically active radiation, wind speed and air specific humidity, temperature, and carbon dioxide concentrations. These profiles, which define the plant stand microclimate, can be determined from a microclimate model.

8.3 Stand Microclimate Models

In homogeneous, closed-canopy, stands of vegetation of large areal extent, most of the variation of the environmental parameters affecting photosynthesis occurs along vertical gradients. These vertical gradients characterize the microclimate of the stands and serve as the pathways along which they exchange energy and mass with the external environment. Horizontal variation under these conditions is small enough that average values of most parameters at each height in the stand are sufficient to describe the local environment. However, the non-linear response of photosynthesis to photosynthetically active radiation requires that the horizontal variation in this factor must be considered when determining photosynthesis at any one canopy level.

A microclimate model, which predicts the vertical gradients and fluxes for the environmental parameters is developed in the following sections. For this development, we have found it convenient to group our discussions into radiative transfer processes and convective transfer processes. However, these processes are closely linked and this separation is used only as a means to facilitate the developments of our model.

8.3.1 Radiative Transfer

The divergence of photosynthetically active and total short-wave radiation flux densities with height are caused by the interception and absorption of this radiation by successive leaf strata in the canopy. Radiation, which is not absorbed by a particular stratum, is transmitted or reflected to other leaves, the sky, the ground, or other plant surfaces.

The pattern of this radiation divergence depends on stand structure, leaf optical properties and angle of incidence of the radiation. Furthermore, the irradiance can be divided into a direct beam component which originates from the very small solid angle occupied by the sun and a diffuse component which comes from all parts of the sky.

Duncan, et al. (1967) have developed an expression for the interception of direct beam radiation for a stand of randomly spaced leaves,

$$\frac{dS}{dz} = \frac{A\,(F/F')}{\sin(k_s)}\,S_o \tag{14}$$

where:

- A = leaf area per increment of height dz
- (F/F') = the apparent leaf area viewed from the angle of incidence over the actual leaf area
- k_s = the solar elevation angle
- S_o = the amount of radiation (photosynthetically active or total short-wave) received on a plane above this increment

This equation can also be used for interception of diffuse sky radiation if it is integrated over all of the angles for the sky hemisphere. The amount of this radiation absorbed depends on the leaf optical properties. However, not all of the radiation which is transmitted or reflected after interception is lost from the canopy. Some of this radiation will be absorbed after multiple reflections and transmissions from the leaves. The modeling procedure mimics this behavior through iterative calculations on values of reflected and transmitted radiation.

This same technique can be used to calculate the average flux divergence of photosynthetically active radiation by using the leaf optical properties in this radiation band. However, as noted before, one needs to calculate photosynthesis at any height for a number of narrow irradiance classes and average the resultant values rather than use an average value of intensity in equation (5). To do this, we have followed the method of DUNCAN, et al. (1967) who assumed that all of the leaves at a particular height were at a given angle to the horizontal and that they were equally distributed with respect to azimuth angle. With these assumptions and given a vertical distribution of leaf area, DUNCAN, et al. could determine the leaf area in direct beam radiation, and the distribution of leaf area at different angles from perpendicular to the direct beam radiation. This enabled him to calculate the amount of leaf area in each of a number of narrow direct beam intensity classes at each height in the canopy. The leaves illuminated by direct beam radiation were assumed to have a total illumination which was the sum of the direct beam in that class and the average diffuse radiation absorbed. The remaining leaf area was assumed to receive only diffuse radiation.

Long-wave radiation is reflected and emitted by the canopy elements as well as intercepted and absorbed by them. The sky and the ground surface are also sources of long-wave radiation. The vegetation is very close to being a black-body radiator in this part of the spectrum. Thus, radiation emitted by leaves will be close to the maximum possible at their absolute temperature and nearly all long-wave radiation intercepted by these surfaces will be absorbed.

Long-wave radiation exchange between canopy layers, the sky, and the ground can be quantitatively described through the use of view factors. The view factor is the line of site view one object has for another, that is, the portion of view from a point on one surface subtended by the other surface. The view factor of a layer of leaves for every other layer, the ground, and the sky is a function of the leaf areas and angles in the layers and the leaf area between the two layers. Reasonable view factors can be determined with the same rationale as that used for the penetration of direct beam photosynthetically active or total shortwave radiation into the canopy.

Once the view factors have been determined for a given canopy geometry, the net long-wave flux for a layer can be described by the following equation from KREITH (1965) page 223.

$$L_n = V_{i-sk} \sigma (T_{sk}^4 - T_i^4) + V_{i-g}\sigma(T_g^4 - T_i^4) + \sum_j V_{i-j} \sigma (T_j^4 - T_i^4), \quad (15)$$

where:

V_{i-sk} = view factor of layer i for the sky,
V_{i-g} = view factor of layer i for the ground,
V_{i-j} = view factor of layer i for layer j,
T = apparent absolute temperature of the sky (sk), absolute ground temperature (g), and absolute leaf temperature in layer j.

8.3.2 CONVECTIVE TRANSFER

The divergences of the short- and long-wave radiation fluxes within the canopy provide the energy exchanged at the individual leaf surfaces which in turn determines the flux densities of the sources and sinks for convective transfer at different levels within the canopy. The actual vertical gradients of air temperature, specific humidity, carbon dioxide concentration, and wind speed which characterize the leaf environment are a result of the interaction between the flux densities of sensible heat, latent heat, carbon dioxide and momentum, and the transport properties of the atmosphere is dominated by convection through the exchange of turbulent eddies.

Above the vegetation, flux densities are constant and eddy size is linearly proportional to the distance above the upper canopy surface. In this region, the eddy diffusivity for momentum is defined by the equation

$$K_m = u_* k(z-d)/\emptyset, \quad (16)$$

where:

K_m = turbulent diffusivity for momentum,
u_* = friction velocity, a convenient grouping of the square root of the shearing stress divided by the density of air,
k = VON KARMAN's constant (0.4),
z = height above the ground,
d = displacement of the wind profile by the stand,
$\emptyset$ = a buoyancy correction to account for atmospheric stability.

The diffusivities for sensible heat and mass follow the same equation; however, the buoyancy correction may have a different value under the same conditions (BUSINGER, et al., 1970).

Inside the stand, the eddy size is related to the size of eddies shed by obstacles such as leaves and branches, and to the distance between these obstacles. No way has been found to directly relate the eddy size to measurable structural characteristics of the stand; however, several investigators have demonstrated that the apparent eddy diffusivities decay exponentially with depth into the canopy (UCHIJIMA & WRIGHT, 1964; WRIGHT & BROWN, 1967).

$$K_i = K_H \exp[B_i(z-H)], \tag{17}$$

where:

K_i = diffusivity for flux density for property i (energy, mass or momentum),
K_H = diffusivity for flux density at the top of the canopy,
B_i = extinction coefficient for apparent eddy diffusivity,
H = height of top of canopy.

The above function for diffusivity can be used in conjunction with flux divergence equations to determine the profiles of the atmospheric fluid properties within the stand.

$$\rho c_p \frac{d}{dz}\left(K \frac{dT_a}{dz}\right) + S_h = 0 \tag{18}$$

$$\rho L_t \frac{d}{dz}\left(K \frac{dq_a}{dz}\right) + S_w = 0 \tag{19}$$

$$\frac{d}{dz}\left(K \frac{dC_a}{dz}\right) + S_c = 0 \tag{20}$$

$$\rho \frac{d}{dz}\left(K_m \frac{du}{dz}\right) + S_m = 0 \tag{21}$$

where:

K = the diffusivity for sensible heat and mass,
S_h = source strength for sensible heat,
S_w = source strength for latent heat,
S_c = source strength for carbon dioxide flux,
S_m = source strength for momentum.

Above the canopy, the source (or sink) strengths for all of the convective fluxes are essentially zero. Within the canopy the source strength for sensible heat and latent heat and the sink strength for carbon dioxide are the flux densities at the leaves. Thus, values for the flux densities at any height can be found by solving equations (5) and (11) with input values of the leaf environment at that height and then multiplying by the leaf area. Since the leaf environment necessary to solve equation (5) includes the air CO_2 concentration found from equation (20) and the leaf environment needed to solve equation (11) includes air temperature and specific humidity found from equations (18) and (19), these sets of equations must be solved simultaneously. While an exact simultaneous solution of these equations is not mathematically possible, an adequate solution can be obtained through the use of iterative computation loops.

8.4 Stand Photosynthesis Model

Taken together, the sets of equations, developed in the pres ceding sections, represent a model for stand photosynthesis. Use of this model to simulate the environmental response of stand photosynthesis requires the simultaneous solution of these equation over a range of environmental boundary conditions. A sufficient set of these boundary conditions includes air temperature, air specific humidity, wind speed and CO_2 concentration at some height above the stand, the incoming short-wave, long-wave and photosynthetically active radiation flux densities above the stand, the shear stress above the stand, and either values of latent heat, sensible heat and CO_2 flux densities or values of specific humidity, air temperature and CO_2 concentration near the soil surface.

Ideally, the lower temperature boundary should be set at the soil depth where the temperature will remain constant at the climatic average. However, because of its large thermal lag, the soil cannot be treated simply as a steady state system. Rigorous evaluation of the soil temperature profile and heat flux densities requires the use of a non-steady state model which utilizes a much more complex set of equations than is necessary to define the atmospheric and vegetation portions of the system. We have avoided this complexity by specifying a slowly changing heat flux lower boundary condition at the soil surface.

The vegetation characteristics required for the simulation are the distributions of leaf area, size and angle with height from the ground, the aerodynamic parameters of zero plane displacement and roughness height, the canopy extinction coefficient for turbulent diffusivity, the leaf optical properties, and the leaf biochemical constants for photosynthesis, respiration and stomatal response.

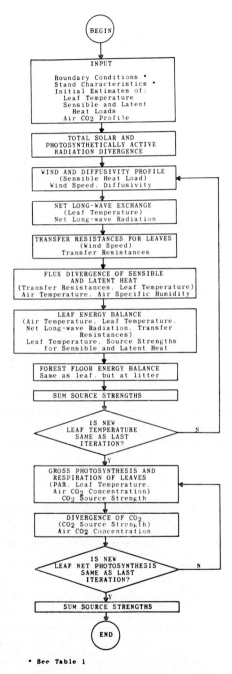

* See Table 1

Figure 1. Solution scheme (model) used for the prediction of stand photosynthesis.

Previous results of stand model simulations have been presented for the energy balance portion of the system by WAGGONER, REIFSNYDER, & FURNIVAL (1968) and MURPHY & KNOERR (1970) and for both energy balance and photosynthesis by WAGGONER (1969), STEWART & LEMON (1969) and SINCLAIR, ALLEN, & STEWART (1971).

The solution scheme used for the current version of the model is illustrated in Figure 1. This scheme is divided into a stand energy balance section and a photosynthetic section. The output of the energy balance section, including the profiles of photosynthetically active radiation, leaf temperature, canopy diffusivities and leaf boundary layer resistances, serves as an input into the photosynthetic section. It is necessary to use iterative routines to solve both sections. A detailed account of the procedures used to solve the individual equations can be found in MURPHY & KNOERR (1970).

8.5 Simulation Results and Discussion

To illustrate the utility of using predictive models, the results of several photosynthesis simulations for a tulip tree stand (*Liriodendron tulipifera* L.) are shown in Figure 2. Two environmental boundary conditions, wind speed and solar radiation, were varied for the purpose of these simulations. The values of the characteristics for the *Liriodendron* stand were taken from the papers of LOACH (1967), EDWARDS & SOLLINS (1972), and WOODS & TURNER (1971). They are listed in Table 1 along with values of the environmental boundary conditions which were not changed during the simulations.

TABLE 1

Selected boundary conditions and stand characteristics*.

	Boundary Conditions		
	Air Temperature	Air Relative Humidity	CO^2
Upper (41 m)	297 °K	0.50	315 vpm
Lower (soil surface)	293 °K	0.55	2 mgm/dm_2 – hr
	Stand Characteristics		
Height	Crown Depth	Zero Plane Displacement	Roughness Length
31.05 m	23 m	29.08 m	17 cm

* Other conditions are available on request.

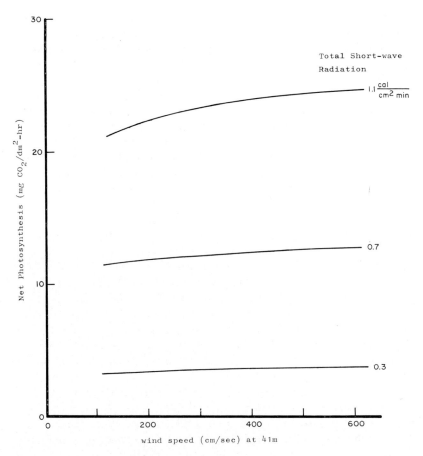

Figure 2. Simulated photosynthetic rates in a stand of *Liriodendron tulipifera* calculated from a predictive model.

The most striking prediction made by the model is that the net photosynthetic rate is relatively insensitive to changes in wind speed while very sensitive to changes in short-wave radiation. This result was not entirely unanticipated since short-wave radiation provides the energy source for photosynthesis. For the purposes of this simulation, the photosynthetically active radiation was assumed to be four-tenths of the total short-wave radiation (PALTRIDGE, 1970).

The lack of large variation in photosynthesis with wind speed is largely a result of the limited effect of wind speed on the flux of CO_2 reaching the chloroplast. Wind speed does affect both the turbulent diffusivities within the stand and the boundary layer resistance close to the leaves. However, diffusion outside of the leaf provides only a small portion of the total resistance in the carbon dioxide path to the chloroplasts and thus has only a limited effect on photosynthesis.

Wind speed also has an effect on respiration through its effect in the energy balance on leaf temperature. Higher wind speeds cause leaf temperature to approach air temperature. For the conditions used in these simulations, this means that leaf temperature will decrease. While the respiration model predicts decreasing respiration and thus increasing net photosynthesis with decreasing leaf temperature, this effect is quite small compared to the direct influence of photosynthetically active radiation on net photosynthesis.

The predicted variations with height for some of the environmental and stand parameters under a high radiation input condition (total solar radiation equals 1.1 cal/cm^2-min) are shown in Figures 3A—3F. These predictions show that most of the photosynthesis takes place near the top of the stand where light and transport conditions are optimum. The model also predicts rather high litter surface temperatures; this is not unusual for conditions at the forest floor. However, the predicted air temperature in the trunk space seems to be somewhat higher than expected; possibly a result of the inadequacy of the turbulent flow model in this region.

In their present form, the several models for stand photosynthesis, including our own, can be solved to predict the response of photosynthesis to a range of environmental and stand parameters. The results of several simulations given above are intended only as an example of what can be done. While such simulations currently give some insight into the important factors controlling photosynthesis, the predictions must ultimately be compared with the real world. Our current research as part of the Eastern Deciduous Biome, U.S. International Biological Programs will emphasize such comparisons. We anticipate that they will provide a basis for testing and improving several of the models. The ultimate objective of this type of research is to provide models that yield predictions realistic enough to form the basis for making decisions related to the management of biological systems.

8.6.1 SUMMARY

This paper outlines the rationale and theory for a model of stand photosynthesis. The development and implementation

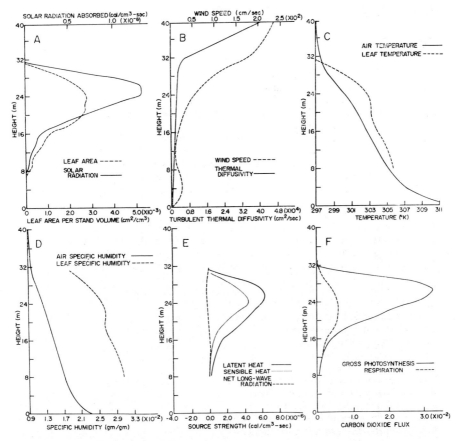

Figure 3A–F. An input profile of leaf area index (A) and model generated profiles of solar radiation (A), wind speed and thermal diffusivity (B), air temperature and leaf temperature (C), air specific humidity and saturation specific humidity at leaf temperature (D), source strengths for latent heat, sensible heat and long-wave radition (E), and source strengths for grass photosynthesis and respiration in mg/dm³-hr (F).

of such a model has been made possible by the historical accumulation of a large amount of information about the mechanisms which control plant processes and the physical parameters of the environment. The development of fast, easily programmed digital computers has also been necessary. We have illustrated how this model can provide predictions of the response of stand photosynthesis to environmental parameters. However, there is still much basic information that must be supplied if these models are to provide predictions useful for the management of biological systems. A major use of the current models will be to guide such research.

8.7 Appendix I

8.7.1 Symbol List

A = leaf area per increment of height
A_c = CO_2 acceptor in the photosynthetic process
A_{cw} = projected area of cell walls per area of leaf
B_i = extinction coefficient for apparent eddy diffusivity of material i
C_a = CO_2 concentration in the air
C_c = CO_2 concentration in the cell at the chloroplast
D = effective leaf width
E = latent heat flux density from evaporation of water
F = tortuosity factor for liquid diffusion through cell walls
F/F' = the apparent leaf area viewed from an angle divided by the actual leaf area
H = sensible heat flux density
I = irradiance of photosynthetically active radiation
I' = constant in the stomatal resistance equation which may be a function of species and preconditioning
K_i = turbulent diffusivity for flux density of property i in the atmosphere
L = long-wave radiation irradiance
L_n = net long-wave radiation absorbed in a layer of the stand
L_t = latent heat of vaporization for water
M = metabolic heat storage from net photosynthesis
P = atmospheric pressure
P_g = gross photosynthesis
P_n = net photosynthesis
Q = sensible heat storage
Q_{10} = rate of change of respiration for a 10° C change in temperature
R = long-wave radiation emittance
R_s = mitochondrial respiration
R_x = base respiration rate at temperature T_x
S = short-wave radiation irradiance
S_i = source strength in the canopy for flux density i
T = absolute temperature (°K)
V_{i-j} = radiant view factor of leaves in strata i for those in strata j
c_p = specific heat of air
d = displacement of the wind profile by the stand
d_w = diffusivity of CO_2 in liquid water
k = von Karman's constant
k_i = chemical rate constants
k_s = solar radiation angle
q_a = specific humidity of the air
q_1 = saturation specific humidity at leaf temperature
r_i = diffusion resistances for heat and mass in different parts of the diffusion path ways
s = solubility of CO_2 in water
t = thickness of a cell wall
u = wind speed
u_* = friction velocity, the square root of the surface shearing stress to atmospheric flow divided by the density of air
z = height above the ground surface

8.7.2 Subscripts

H   = at canopy height
L   = for long-wave radiation
S   = for short-wave radiation
a   = air
cw = cell wall
g   = ground
h   = heat
l   = lower
m = momentum
p   = protoplasmic
s   = stomatal
sk = sky
u   = upper
w   = latent heat or water

8.7.3 Greek Letters

α = stress factor for stomatal resistance, dependent on temperature and water stress
α_i = absorptivity for radiation in band i
β = constant in the stomatal resistance equation, function of species and preconditioning
$\emptyset$ = bouyancy correction to account for atmospheric stability in the diffusivity and wind equations
ρ = density of air
σ = Stefan-Boltzmann constant

REFERENCES

Ashby, W. R. – 1956 – An Introduction to Cybernetics. Chapman and Hall Ltd. London, 295 pp.

von Bertalanffy, L. – 1968 – General System Theory. George Braziller, Inc. New York, 289 pp.

Businger, J. A., J. C. Wyngaard, Y. Izumi & E. R. Bradley – 1971 – Flux-profile relationships in the atmospheric surface layer. *J. Atmos. Sci.* 28: 181–189.

Duncan, W. G., R. S. Loomis, W. A. Williams & R. Hanau – 1967 – A model for simulating photosynthesis in plant communities. *Hilgardia*, 38, No. 4, pp. 181–205.

Edwards, N. T. & P. Sollins – 1973 – Continuous measurement of carbon dioxide evolution from partitioned forest floor components. *Ecology* 54: 406–412.

Gaastra, P. – 1959 – Photosynthesis of crop plants as influenced by light, carbon dioxide, temperature and stomatal diffusion resistance. *Meded. Landb. Hogesch. Wageningen* 58, 1–68.

Gates, D. M. – 1968 – Transpiration and leaf temperature. *Ann. Review Plant Phys.* 19: 211–238.

HALL, A. E. – 1971 – A model of leaf photosynthesis and respiration. Annual Rept. of Dir. Dept. of Plant Biology Carnegie Institution, Stanford, California. 530–540.
KNOERR, K. R. & L. W. GAY – 1965 – Tree leaf energy budget. *Ecology* 46: 17–24.
KREITH, F. – 1965 – Principles of Heat Transfer. International Textbook Co., Scranton, Penn. 620 p.
LOACH, K. – 1967 – Shade tolerance in tree seedlings. *New Phytol.* 66 (4): 607–621.
LOMMEN, R. W., C. R. SCHWINTZER, C. S. YOCUM & D. M. GATES – 1971 – A model describing photosynthesis in terms of gas diffusion and enzyme kinetics. *Planta* 98: 195–220.
MARTIN, F. F. – 1968 – Computer Modeling and Simulation. John Wiley & Sons, New York, 331 p.
MURPHY, C. E. & K. R. KNOERR – 1970 – A general model for energy exchange and microclimate of plant communities. in Proceedings, 1970 Summer Computer Simulation Conference, ACM/SHARE/SCI, Denver, Colorado, p. 786.
MURRAY, F. W. – 1967 – On the computation of saturation vapor pressure. *J. Applied Met.* 6 (1): 203–204.
PALTRIDGE, G. W. – 1970 – A filter for absorbing photosyntheticallyactive radiation and examples of its use. *Ag. Meteor.* 7 (2), 167–174.
PARKHURST, D. F., P. R. DUNCAN, D. M. GATES & F. KREITH – 1968 – Wind-tunnel modelling of convection of heat between air and broad leaves of plants. *Agr. Meteorology* 5: 33–47.
PATTEN, B. C. – 1971 – Systems Analysis and Simulation in Ecology, Volume 1. Academic Press, New York, 2560 p.
RABINOWITCH, E. I. – 1951 – Photosynthesis and Related Processes Volume II. Parts 1 & 2. Interscience Publishers Inc., New York.
RASCHKE, K. – 1956 – Über die physikalischen Beziehungen zwischen Wärmeübergangszahl, Strahlungsaustausch, Temperatur und Transpiration eines Blattes. *Planta* 48: 200–238.
SHAWCROFT, R. W. – 1970 – Water relations and stomatal response in corn. Ph. D. Thesis, Cornell University, Ithaca, New York.
SINCLAIR, T. R., L. H. ALLEN & D. W. STEWART – 1971 – A simulation model for crop-environmental interactions and its use in improving crop productivity. *Proceedings*: *1971 Summer Computer Simulation Conference, Boston, Mass., Board of Simulation Conference, Denver, Colorado*, 784–794.
SINCLAIR, T. R. – 1972 – A leaf photosynthesis submodel for use in general growth models. *U.S.I.B.P., Triangle Research Site, Eastern Deciduous Forest Biome, Memo Rpt. 72–14, Durham, N.C.*, 14 p.
STEWART, D. W. & E. R. LEMON – 1969 – A simulation of net photosynthesis of field corn. *Microclimate Investigations Interim Rpt. 69–3, U.S.D.A., Cornell Univ.* Ithaca, New York.
STEWART, D. W. – 1970 – A similation of net photosynthesis of field corn. Ph. D. Thesis, Cornell University, Ithaca, New York.
THOM, A. S. – 1968 – The exchange of momentum, mass, and heat between an artificial leaf and the airflow in a wind-tunnel. *Quart. J. Royal Met. Soc.*, 94: 44–55.
UCHIJIMA, A. & J. L. WRIGHT – 1964 – An experimental study of air flow in a corn-air layer. *Bull. Nat. Inst. Agr. Sci. (Japan), Series A*, No., 11, pp. 19–65.
WAGGONER, P. E. – 1969a – Predicting the effect upon photosynthesis of changes of leaf metabolism and physics. *Crop. Sci.* 9: 315–321.
WAGGONER, P. E. – 1969b – Environmental manipulation for higher yields.

in: Physiological Aspects of Crop Yield. ed. EASTIN, HASKINS, SULLIVAN, VAN BAVEL, ASA, CSSA: 343–374.

WAGGONER, P. E., G. V. FURNIVAL & W. E. REIFSNYDER – 1969 – Simulation of the microclimate in a forest. *Forest Sci.* 15: 37–45.

WOODS, D. B. & N. C. TURNER – 1971 – Stomatal response to changing light by four tree species of varying shade tolerance. *New Phytol.* 70: 77–84.

WRIGHT, J. L. & K. W. BROWN – 1967 – The energy budget at the earth's surface: comparison of momentum and energy balance methods of computing vertical transfer within a crop. *Tech. Report ECOM 2-67 1-1*, Fort Huachuca, Arizona.

9 EXPERIMENTAL ANALYSIS OF ECOSYSTEMS

J. FRANK MCCORMICK, ARIEL E. LUGO & REBECCA R. SHARITZ

Contents

9.1	Introduction	151
9.1.1	The Ecosystem Concept	151
9.1.2	The Experimental System	152
9.2	Experimental Design	154
9.3	Experimental Analysis of Ecosystem Structure	155
9.3.1	Descriptive Field Studies	155
9.3.2	Development of Hypotheses	160
9.3.3	Test for Cause and Effect Relationships	160
9.3.4	Laboratory Experiments	160
9.3.5	Field Experiments	166
9.4	Experimental Analysis of Ecosystem Function	167
9.4.1	Species Metabolism	179
9.4.2	Ecosystem Metabolism	171
9.5	Relationships of Ecosystem Structure and Function	173

9 EXPERIMENTAL ANALYSIS OF ECOSYSTEMS

9.1 Introduction

9.1.1 The Ecosystem Concept

The ecosystem is both a useful concept and a physical reality. Ecosystems are dynamic units of nature possessing structural and functional characteristics which vary in magnitude or rate within the environmental dimensions of space and time. A chief goal of ecology is to identify and quantify relationships between structure and function in these definable units of nature. The objective of this chapter is to describe experimental approaches to ecosystem analysis.

The biological components of ecosystems are populations of species which share a common environment. Populations within an ecosystem interrelate and interact through their coordinated roles in the flow of energy, the cycling of elements, and modification of microenvironments within the system. Since population attributes such as density, distribution, growth and reproduction fluctuate within the dimensions of space and time, descriptions of ecosystem structure and function should include these fluctuations. Integrated measures of ecosystem structure or metabolism may not always reflect fluctuations in individual species because of compensation or synergism within the system. Such interactions are, in fact, responsible for many unique qualities of ecosystems. However, fluctuations in individual components of the system should be known in order to design metabolic studies, to interpret results, and to understand relationships between structure and function.

Studies of small terrestrial ecosystems which occur on rock outcrops in the southeastern United States have been used to demonstrate experimental approaches to ecosystem analysis. It is intended that studies of these relatively simple ecosystems will serve as models for experimental analysis of other ecosystems. Quantitative analyses of the population dynamics of dominant plant species provide descriptions of the structural components of the system and identify the factors which most strongly control structure. Experimentation begins with field observations which lead to the development of hypotheses concerning ecosystem structure and metabolism, as well as species and system adaptations to environmental stresses. These hypotheses can be tested under controlled laboratory conditions and, finally, cause and effect relationships are verified under modified natural conditions in a field laboratory.

9.1.2   THE EXPERIMENTAL SYSTEM

Many natural ecosystems are so large and complex that experimental manipulation is difficult. Also, the number of environmental variables and species interactions in a diverse system may defy analysis of cause and effect relationships. Successful studies of ecosystem structure and function have most often involved relatively small and simple ecosystems, such as ponds. Ideally, an experimental ecosystem should have clearly defined boundaries; it should be small enough to manipulate; and it should contain large numbers of relatively small organisms so that the demands of sampling and statistical analysis are satisfied. Also there should be limited immigration or emigration; the system should be relatively closed; and many replicates should be available. Small terrestrial ecosystems which occur in depressions on the granite outcrops satisfy these requirements. These granite outcrops extend along the contact between the coastal plain and piedmont from northeastern North Carolina to eastern Alabama. Similar rock outcrops of various parent materials occur throughout the world and support similar ecosystems. Individual bowl-shaped depressions develop as surface rock weathers, usually along drainage paths. Depressions usually vary from 2 to 4 meters in diameter; they accumulate weathered granite and wind-blown debris to a depth of 20 to 50 centimeters. These manipulatable ecosystems are analogous to terrestrial islands floating on a sea of granite. They are exposed to severe environmental stress due to high light intensity, extreme and rapid fluctuations of moisture and temperature, and shallow, acid, sandy soils (Figure 1).

Another advantageous feature is that these simple "island" ecosystems contain no animals larger than insects. Even insects, however, are not abundant and grazing is minimal. The chief role of the microfauna is the decomposition of dead organic matter. Therefore, almost all structure and metabolism in the system is attributable to vegetation. The single but significant exception is soil respiration which, under high temperature and moisture conditions, is considerable (LUGO, 1969).

Figure 1. "Island" ecosystem on Mt. Arabia, Georgia, showing conspicuous zonation of *Sedum smallii* (arrow 1) and *Minuartia uniflora* (arrow 2). The location of *Viguiera porteri* is indicated by the dead stalks of the previous year's plants (arrow 3) and the current year's seedlings beneath them. The island ecosystem is surrounded by the lichen-covered granite outcrop.

Floristic studies (McVaugh, 1943; Burbanck & Platt, 1964) list 250 species which may be found on the granite outcrops but only about 40 are characteristic. Of these, 14 are endemic to these particular rock outcrops. This is indicative of the unique environmental conditions and evolutionary history of the flora. Of the approximately 40 characteristic species, fewer than 10 contribute significantly to the structure and metabolism of these miniature "island" ecosystems.

Sedum smallii (Britt.) Ahles, *Minuartia uniflora* (Walt.) Mattf. and *Viguiera porteri* (A. Gray) Blake are of greatest significance on the basis of relative density, frequency, and biomass. Each occupies a distinct habitat and, collectively, these three species account for most of the productivity in the herbaceous annual ecosystems.

Sedum germinates as early as October, overwinters as a rosette, begins to grow again in March, and flowers in April. *Sedum* occupies the most shallow (0—4 cm) peripheral soils of the "island" ecosystem. *Minuartia* has a similar life cycle and occurs as a zone of vegetation just interior (4—12 cm) to that of *Sedum*.

The shallow peripheral soils occupied by these plants during the winter and spring are too dry to support significant plant populations during the summer. In deeper soils (> 12 cm) toward the interior of the "island", dense populations of *Viguiera* completely dominate the ecosystem. Seeds of *Viguiera* germinate early in the spring, the seedlings grow to maturity throughout the summer, and flower in the autumn. If soil accumulation exceeds a depth of 20 cm, *Viguiera* may be replaced by the perennial *Senecio tomentosus* Michx.

9.2 Experimental Design

An important aspect of experimental design is to select an experimental ecosystem which can be manipulated in the manner which one's research objectives dictate. The larger and more complex the ecosystem, the more complex the technology required for analysis. However, with time, patience, money, and hard work, experimental analysis of any ecosystem is possible as evidenced by the giant metabolism chamber constructed in the tropical rain forest at El Verde, Puerto Rico by ODUM (1970).

Experimental ecosystem analysis should proceed from the most simple and descriptive to the most complex and experimental. Ideally, such analysis should proceed from field observations to laboratory experiments, and finally to the modification of natural conditions in a field laboratory (PLATT & GRIFFITHS, 1964) to determine whether relationships identified in the laboratory actually exist in the field. The following outline serves as a procedural guide for experimental analysis of ecosystem structure and function. The upper and lower lines identify specific topics which, at each stage of analysis, should be investigated in order to understand ecosystem structure (upper) or function (lower).

Structure	Population Dynamics	Controls over Structure	Identify Cause & Effect Relationships	Test Tolerance Levels & Relationships in the Field	
ANALYTICAL SEQUENCE	FIELD OBSERVATIONS	→DEVELOPMENT OF HYPOTHESES	→LABORATORY EXPERIMENTS	→FIELD LABORATORY	Interpret →Relations of Ecosystem Structure to Function
Function	Environmental Analysis	Adaptations to Environment	Species Metabolism	Ecosystem Metabolism	

TABLE 1

Reproductive potential for *Sedum smallii* and *Minuartia uniflora* on Mt. Arabia

Species	Seeds per flower	Flowers per plant	Seeds per plant	Plants per dm^2	Seeds per dm^2
Sedum smallii	12.0	9.5*	114.1	41.2	4698
Minuartia uniflora	12.5	24.4	305.2	9.6	2946

* From WIGGS & PLATT, 1962.

9.3 Experimental Analysis of Ecosystem Structure

9.3.1 Descriptive Field Studies

Ecosystem structure is usually determined by analyzing the distribution, abundance, and biomass of the component species populations. Observations by many investigators and systematic descriptions by a few (BURBANCK & PLATT, 1964; McCORMICK & PLATT, 1962) suggest that habitat specificity in the "island" ecosystems is due to the influence of soil depth and moisture upon the competitive ability of species which dominate specific habitats. In order to test these assumptions and to obtain quantitative descriptions of ecosystem structure, field sampling studies were initiated to measure fluctuations in species density, distribution, biomass, and reproduction. Ecosystems were sampled throughout the year (SHARITZ & McCORMICK, 1973; MELLINGER, 1971). Simultaneous measurements were made of precipitation and also of soil moisture at various depths. Solar radiation, substrate and air temperatures, carbon dioxide concentrations, and relative humidity were also measured at intervals (LUGO, 1969).

Results of systematic sampling (Figure 2) supplemented with results of germination studies (SHARITZ & McCORMICK, 1973; MELLINGER, 1971) provide the data necessary to calculate reproductive potential (Table 1), to construct actuarial life tables (Table 2) and survivorship curves (Figure 3) for the most important components of the system.

As soil depth increases from the periphery toward the center of the ecosystem, there is a linear increase in soil moisture and biomass (SHARITZ & McCORMICK, 1973). Daily fluctuations in temperature and carbon dioxide as well as the generally high solar radiation and low moisture availability indicate that the factors which most strongly influence primary production fluctuate rapidly

TABLE 2

Life table for a natural population of *Sedum smallii*

x	D_x	A_x	A'_x	l_x	d_x	$1000q_x$	L_x	T_x	e_x
Seed produced	4	0– 4	−100.0	1000	160	160.00	920.0	4435.5	4.43
Available	1	4– 5	− 9.7	840	630	750.00	525.0	755.5	0.90
Germinated	1	5– 6	+ 12.9	210	177	842.86	121.5	230.5	1.10
Established	2	6– 8	+ 35.4	33	9	272.72	38.5	109.0	3.30
Rosettes	2	8–10	+ 80.6	24	10	416.67	19.0	52.0	2.17
Mature plants	2	10–12	+125.7	14	14	1000.00	7.0	14.0	1.00

Life cycle stage   x

Duration of stage D_x = number of months of each stage

Age A_x = population age in months

Percentage Age A'_x = population age at beginning of each stage as percent of mean life span

Survivorship l_x = number* of individuals at beginning of each stage

Senescence d_x = number* of individuals which die during each stage

Mortality rate $q_x = d_x/l_x$

Stationary Population $L_x = (l_x + l_{x+1})/2$ $T_x = D_x L_x + T_{x+1}$

Residual Population Life Span

Life Expectancy $e_x = T_x/l_x$

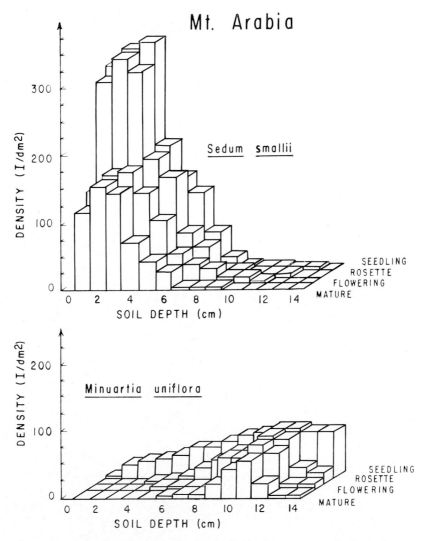

Figure 2. Histogram of species density (individuals per square decimeter) as a function of soil depth (±0.5 cm) and life cycle stage in island communities on Mt. Arabia, Georgia. Each bar represents a mean density of all plots sampled under the designated conditions.

and frequently reach limiting intensities (SHARITZ & MCCORMICK, 1973; LUGO, 1969).

When species density (Figure 2), biomass (Figure 4), or seed production are plotted against an intensity gradient of one or more environmental factors, for example, soil depth, the conditions or locations where critical interactions occur are shown. If these data,

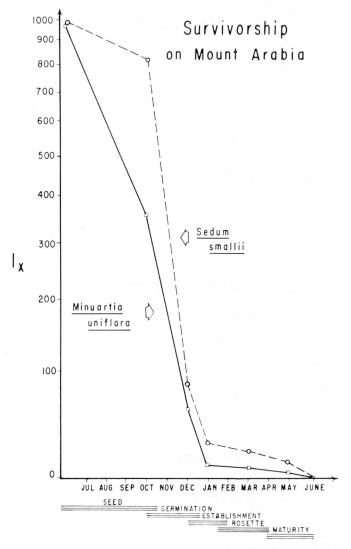

Figure 3. Survivorship (1_x) of natural populations of *Sedum smallii* and *Minuartia uniflora* on Mt. Arabia, Georgia.

or survivorship curves (Figure 3) are plotted against time, one can find the times when the most important interactions occur. Aided with information as to when, where and under what conditions the most critical events occur, hypotheses can be developed concerning the influences of density-dependent and density-independent factors upon ecosystem structure. This information also enhances the experimental design of metabolic studies.

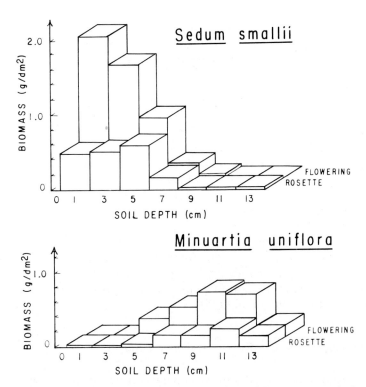

Figure 4. Histogram of species biomass (grams dry weight per square decimeter) as a function of soil depth (± 0.5 cm) and life cycle stage in island communities on Mt. Arabia, Georgia. Each bar represents a mean of four samples.

These field studies have demonstrated the feasibility and value of using quantitative descriptions of the population dynamics of plant species to describe ecosystem structure. Results have provided a rational basis for developing hypotheses concerning the factors which control ecosystem structure. These descriptions of ecosystem structure include the fluctuations in density, distribution, biomass and reproduction which are in part responsible for the dynamic nature of ecosystem structure and function.

Results of these descriptive field studies and observations justify the following hypotheses: (1) Soil depth and moisture are related in a direct linear manner; (2) The three dominant species of the "island" ecosystems, *Sedum smallii*, *Minuartia uniflora*, and *Viguiera porteri*, occur in distinct zones which differ in soil depth and moisture.

A word of caution is appropriate at this point. Correlations between species behavior, ecosystem structure, and environmental factors, when based upon field sampling, do not prove that any of the three are related to one another. These correlations, however, do provide a rational basis for the development of hypotheses which should then be challenged by tests for cause and effect relationships.

9.3.2 DEVELOPMENT OF HYPOTHESES

On the basis of the relationships observed between the population dynamics of component species and intensity gradients of soil depth and moisture, it is proposed that the structure of the "island" ecosystems is most strongly influenced by soil depth, soil moisture, and interspecific competition.

On the basis of field measurements of extreme intensities and rapid rates of change of several environmental factors (LUGO, 1969), it is proposed that the component species of the "island" ecosystems are physiologically adapted to conditions of high light intensity, extreme and rapidly fluctuating substrate temperatures, and limited moisture availability.

9.3.3 TESTS FOR CAUSE AND EFFECT RELATIONSHIPS

A multifactorial laboratory experiment tested species responses to various levels and combinations of competition, soil moisture, and soil depth while attempting to control the influence of other variables. A field study attempted to randomize, rather than control, the influence of all variables except soil depth, soil moisture, and biotic competition.

9.3.4 LABORATORY EXPERIMENTS

A controlled environment chamber was programmed to simulate conditions previously recorded (WIGGS & PLATT, 1956; MCCORMICK, 1959) on granite outcrops. *Sedum*, *Minuartia*, and *Viguiera* seedlings were grown to maturity under 36 combinations of intraspecific competition, interspecific competition, soil depth, and soil moisture with four replications of each set of conditions (SHARITZ & MCCORMICK, 1973).

In these studies, *Sedum* survived in highest densities (Figure 5) and exhibited maximum growth (Figure 6) and seed production (Figure 7) in deep well-watered soil free of interspecific competition. In natural situations, this potential is experienced only in permanent pools on the summits of dome-shaped outcrops where

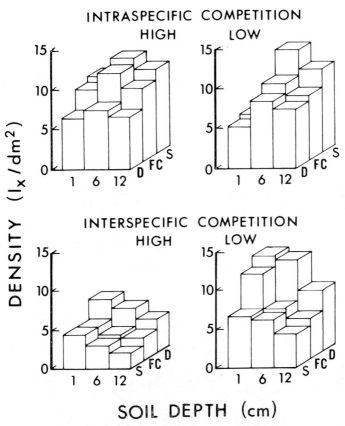

Figure 5. Effect of competition on the density (expressed as a percentage of the initial cohort sown) of *Sedum smallii* at maturity as a function of soil depth and moisture. Treatments are three soil depths (1, 6, and 12 cm) and three relative moisture levels (saturation, S; field capacity, FC; 1/3 field capacity, D).

Sedum occurs in pure populations. The presence of a well-adapted competitor (*Minuartia*) reduces the *Sedum* population and stunts the size of survivors (Figures 5 and 6). Biomass, density, and seed production are inversely proportional to the density of the competitor. As a consequence, in the presence of a competitor, maximum biomass and density occur in the most shallow and dry soil. In natural situations, *S. smallii* is most successful in the driest and most shallow soils where competitors are excluded. Intraspecific competition does not reduce density or biomass as severely as does interspecific competition. Dry, shallow soil does not greatly reduce

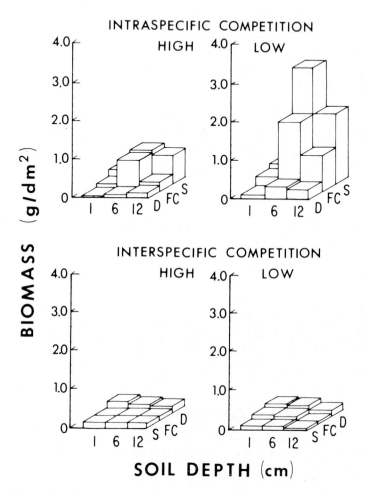

Figure 6. Effect of competition on the biomass (grams dry weight per dm²) of *Sedum smallii* at maturity. Refer to Fig. 5 for treatment identification.

the density of *Sedum*, although plant size and seed production were reduced.

Minuartia also attained highest densities, biomass, and seed production in deep well-watered soil (SHARITZ & MCCORMICK, 1973). *Minuartia* survived in shallow soil if high moisture levels were maintained, however biomass was greatly reduced. As in nature, under dry conditions in shallow soil, no *Minuartia* plants

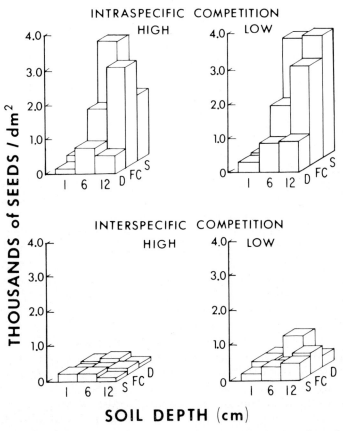

Figure 7. Effect of competition on the seed production (number of seeds in a dm^2) of *Sedum smallii*. Refer to Fig. 5 for treatment identification.

survived. No seeds were produced in shallow soil regardless of the water level applied. The addition of a competitor (*Viguiera*) reduced density, biomass, and seed production under optimal conditions of soil depth and moisture. As a consequence, *Minuartia* survived and grew best in soils of intermediate depth and moisture, just as in natural ecosystems. *Minuartia* gains competitive advantage over *Sedum* as environmental stress decreases. As environmental conditions, specifically soil depth and water, continue to improve *Viguiera* gains in competitive advantage over *Minuartia*. The apparent "optimal" environment for *Sedum* or *Minuartia* is not the set of physical conditions under which the species grows best but the best conditions which are also restrictive to the major competitor. Intraspecific competition also reduces density and biomass of *Minuartia* throughout the range of conditions typical of the *Minuartia* habitat.

Laboratory studies of *Viguiera* water relations, germination requirements, and substrate requirements (MELLINGER, 1972) help to explain the distribution of *Viguiera* within the "island" ecosystems. The combination of conditions necessary for germination of *Viguiera* occurs only in the deepest soils (> 12 cm) or in deep *Polytrichum* mats. These conditions are: (1) ground cover suitable for seed retention but which does not allow burial of seed; (2) a supply of water at high water potentials to the soil surface for 10—12 days during the germination period; (3) substrate pH of 4.0—5.0; (4) seed stratification conditions of high moisture and temperatures below 7°C; (5) alternating day-night temperatures of approximately 20°C and 10°C respectively which allows initiation of germination (MELLINGER, 1972).

Water accumulates in the center of the ecosystems where the depression in the granite substrate is deepest. The perched water table and the high water potentials characteristic of sandy outcrop soils allow *Viguiera* to absorb water at relatively high water potentials during periods of drought. These conditions do not prevail at soil depths less than 12 cm. A favorable soil-to-plant water gradient exists at soil moisture levels above 10 % (MELLINGER, 1972). During a drought the plant-soil water relationship changes within 2—3 days. When *Viguiera* wilts it can remain wilted for as long as 14 days and still recover. No other outcrop species is able to recover after having been wilted so long. Only in soils 12 cm or more in depth do the combinations of conditions exist which favor seed retention, seed germination, and adequate moisture availability for growth and survival of *Viguiera*. In moss mats and in litter, which accumulates near the center of the "island" ecosystems, dormant seeds are retained for one or more years. These seed reservoirs prevent extinction of *Viguiera* during years of no seed production.

Plant responses confirm the hypothesis that soil depth, soil moisture, and competition most strongly influence the growth, survival and distribution of dominant species, and therefore the structure of the ecosystem. More importantly it appears that each species dominates a specific zone or habitat because the local conditions of soil depth and moisture provide the closest thing to "optimal physical conditions" which also restrict other species which might, under more favorable conditions, outcompete the species in question.

Figure 8. Simulated outcrops in the North Carolina Botanical Garden were constructed of concrete (A) which was covered with hot tar (B) and then coated with powdered granite (C) of the same composition as the natural outcrops. The depressions are 2 m in diameter and slope to a depth of 20 cm.

9.3.5 FIELD EXPERIMENTS

A procedural objective of the laboratory studies was to control variations in all factors except those being tested. This is extremely difficult to do in the field. In contrast, we attempted to randomize the influence of all variables, and then superimpose upon an otherwise homogeneous test system, an intensity gradient of one or more factors. Entire island ecosystems were transplanted to simulated rock outcrops in the North Carolina Botanical Garden according to the techniques of PLATT & McCORMICK (1964) (Figure 8).

Using additional ecosystems, top soil containing propagules of all species, was separated from subsoil. Both top soil and bottom soil samples were thoroughly mixed and then first bottom soil, and finally top soil were spread evenly throughout two simulated outcrops. Mixing the soil randomized all variables except soil depth which increased gradually from the outer edges of the simulated ecosystem to a maximum of 20 cm in the center. Propagules of all species had an opportunity to occur with equal frequency throughout the ecosystem. Deviations from random distribution over time showed which life cycle stage is most sensitive to environmental selection along the intensity gradient of the experimental variables. Since one objective was to test the independent effects of soil depth and soil moisture, additional soil samples were similarly mixed and transplanted to another simulated outcrop inclined at an angle of 8°. Drainage along the incline produced a moisture gradient, in effect, perpendicular to the gradient of soil depth. Dynamics of the moisture regime was monitored in three ways: measurement of rainfall, gravimetric determination of percentages of moisture in soil; and conductometric measurements using Bouyoucos blocks. The influence of soil depth and moisture was evaluated by observing the differential survival of all species throughout the system much as one would observe the sorting out of pigments on a two dimensional chromatograph. Species density, distribution, growth and reproduction, as well as soil moisture and precipitation, were measured in undisturbed natural ecosystems and in the experimental ecosystems.

Transplanting natural ecosystems to the simulated outcrops had no adverse effects upon the experimental systems. The simulated outcrops did produce the soil depth and soil moisture gradients desired for experimental analysis (CUMMING, 1969).

Plant populations respond to stress by reductions in size (stunting) and/or reductions in numbers (thinning). *Viguiera porteri* illustrated the stunting response to limited soil depth or moisture. Plant height, number of branches and number of leaves per plant varied directly with soil depth both in May and at the end of the growing season (CUMMING, 1969). The thinning response was equally significant. The sequence of species which occupied successive zones of increasing soil depth was *Talinum teretifolum* Pursh, *Cyperus granitophilus* McVaugh, *Crotonopsis elliptica* Willd., *Viguiera*, and *Senecio tomentosus*. This is precisely the sequence found in natural and control ecosystems.

Differential survival of the several species along the soil depth and moisture gradients resulted in a zonation of species nearly identical to that before transplantation and similar to that of undisturbed ecosystems in their natural environment.

Differential responses in terms of growth and survival of each species along intensity gradients of two co-variables (soil depth and moisture) resulted in patterns of distribution and abundance similar to patterns in natural ecosystems. Each species dominates in a specific zone because of its competitive advantage under the conditions of soil depth and moisture which prevail in that habitat. It is important to understand that these conditions seldom if ever represent the optimal physical conditions for survival, growth, or reproduction of that species. The conditions are the most favorable available which will not support a better competitor. Species zonation along the intensity gradients is a consequence of shifting competitive superiority as physical conditions shift. It is this type of natural selection which restored normal structure in the experimental ecosystems within one year. This structure was also characteristic during the following year. Similar experiments using seedling populations of *Sedum smallii*, *Minuartia uniflora*, and *Viguiera porteri* were conducted by SHARITZ & McCORMICK (1973) and yielded similar results.

In summary, differential responses of dominant species to intensity gradients of soil depth and soil moisture, under field conditions provide a final quantitative confirmation of the hypothesis that the influence of soil depth and moisture upon interspecific competition most strongly controls structure in this particular type of ecosystem.

9.4 Experimental Analysis of Ecosystem Function

Photosynthesis, respiration, and transpiration rates of individual species and entire ecosystems were analyzed using gas exchange techniques (LUGO, 1969; MURPHY & McCORMICK, 1971) under controlled laboratory conditions as well as in the field. In order to relate rates of processes to environmental conditions, and

TABLE 3

Summary of evapo-transpiration and transpiration data

	Number of diurnal determinations	Average transpiration ** g H$_2$O.m^{-2}.day^{-1}	Maximum rate observed g H$_2$O.m^{-2}.day^{-1}	Average rate observed g H$_2$O.m^{-2}.day^{-1}
		Transpiration:		
Viguiera porteri	3	1134.80	179.0	48.0
Senecio tomentosus	6	938.92	282.0	40.2
*Talinum teretifolium**	1	114.84	13.1	4.9
*Bulbostylis capillaris**	1	15.48	4.1	0.72
		Evapo-transpiration:		
Polytrichum commune	4	157.36	33.7	7.6
Selaginella rupestris	1	136.76	16.1	6.4
Soil	2	696.70	95.0	36.6

* Rates expressed in gH$_2$O . gdw^{-1} . time^{-1}.
** All expressions of area refer to area of leaf surface when reporting values for species and area of ground surface when reporting values for ecosystems.

TABLE 4

Summary of metabolic data obtained at the N.C. Botanical Garden under natural conditions**

Species	Month	Number of determinations	Total net Photosynthesis $gC \cdot m^{-2} \cdot day^{-1}$	Total night Respiration $gC \cdot m^{-2} \cdot night^{-1}$	P/R	Maximum P $gC \cdot m^{-2} \cdot hr^{-1}$	Maximum Rni $gC \cdot m^{-2} \cdot hr^{-1}$	Average P $gC \cdot m^{-2} \cdot hr^{-1}$	Average R $gC \cdot m^{-2} \cdot hr^{-1}$
			Individual plant determinations:						
Viguiera porteri	May	6	2.260	2.374	0.95	0.313	0.373	0.190	0.190
	June	3	2.682	2.252	1.19	0.380	0.412	0.218	0.188
	July	1	1.400	4.480	0.31	0.262	0.843	0.116	0.372
Senecio tomentosus	May	8	4.128	3.154	1.30	0.588	0.506	0.340	0.258
	June	3	1.472	5.312	0.27	0.284	0.733	0.164	0.346
*Hypericum gentianoides**	June	4	0.108	0.188	0.57	0.054	0.044	0.006	0.014
*Talinum teretifolium**	June	3	0.008	0.008	1.00	0.003	0.004	0.0006	0.0006
*Coreopsis grandiflora**	May	3	0.016	0.030	0.55	0.005	0.005	0.001	0.002
			Determinations of soil and plants with soil						
Soil	May	2	—	8.918	—	—	0.738	—	0.370
	June	2	—	8.032	—	—	0.823	—	0.344
	August	2	—	13.150	—	—	1.274	—	0.546
	September	2	0.314	1.602	—	0.175	0.280	0.038	0.198
	Averages			7.924			0.778		0.364
Edge of Island Ecosystem	May	1	1.246	2.624	0.47	0.177	0.524	0.026	0.094
Entire Island Ecosystem	June	1	14.30	20.40	0.68	4.00	6.00	1.300	1.568

* Rates presented in $gC \cdot gdry\ weight^{-1} \cdot time^{-1}$.
** All expressions of area refer to area of leaf surface when reporting values for species and area of ground surface when reporting values for ecosystems.

also to obtain additional information necessary to construct energy and water budgets for the ecosystem, temporal and spatial variations were monitored for solar energy, temperature, atmospheric and soil moisture, and carbon dioxide. Biomass, calorific values, leaf area and chlorophyll contents were determined for each species and for each ecosystem being studied (LUGO, 1969).

9.4.1 SPECIES METABOLISM

Population samples of each species were placed in Plexiglas chambers inside a programmed environment room and also under field conditions. Fluctuations were measured in atmospheric carbon dioxide and moisture content of chamber and ambient air (LUGO, 1969). The gas exchange analysis system is described by LUGO (1969).

The transpiration pattern of *Senecio tomentosus* is typical of that of most species under field or laboratory conditions. Transpiration rates decreased during the night when relative humidity was high and temperatures were low (LUGO, 1969). Some water absorption was observed in the early morning. High rates were observed during the day with a temporary depression at midday. Depression in transpiration corresponds to a decrease in light intensity.

The pattern of CO_2 exchange of *Viguiera* is typical of heliophytes which occur in the outcrop ecosystem. Tables 3 and 4 summarize metabolic data which are useful in estimating the metabolism of each species in the experimental ecosystem. LUGO (1969) reported that patterns of CO_2 exchange varied between vascular and non-vascular plants, stages in the life cycle, sunny vs. cloudy days, and between moist and dry conditions.

Vascular plants consistently showed rates of net photosynthesis (Pn) and night respiration (Rni) higher than those of non-vascular plants.

Young seedlings of annual species showed high rates of net photosynthesis during the day with low night respiration. Average net photosynthesis rates were 4.8 times the average night respiration rates. Adult individuals of the same species showed high net photosynthesis rates but night respiration rates were higher than those observed for seedlings. High photosynthetic rates were maintained at high light intensities suggesting high light adaptation.

During sunny days vascular plant Pn rates were high for long periods of time, and P/R ratios were as high as 2.4. Cloudy days resulted in low Pn rates and P/R ratios as low as 0.31. These ratios also fell below one when moisture became limiting.

Maximum rates of both water and CO_2 exchange were observed in the early morning and maximum Rni was observed in the early evening. These patterns may be due to environmental conditions prevailing at particular times. Early in the morning moisture conditions are favorable in plant tissue and in the soil; ambient CO_2 is high, light and temperature are increasing, and tissue sugar concentrations are low. All these conditions favor high rates of Pn. After sunset, temperatures are still high and tissue sugar content is high; both conditions favor high Rni.

Results of species metabolism studies indicate species adaptations to conditions of high moisture stress and high light intensities characteristic of the outcrop ecosystem environment (BURBANCK & PLATT, 1964). Outcrop species demonstrate two methods of adaptive strategy (LUGO, 1969). One group of non-vascular species and a few vascular perennials (*Polytrichum commune* Hedw., *Selaginella rupestris* (L.), and *Senecio*) exhibit slow growth rates over extended periods of time and withstand desiccation for long periods of time. The other group of annual vascular plants, including *Viguiera, Hypericum gentianoides* (L.) BSP and *Crotonopsis* exhibit explosive growth rates when moisture conditions are favorable but are able to withstand desiccation for much shorter periods of time. Metabolism in both groups is most severely limited by available soil moisture and both groups are adapted to high light intensities. The latter is reflected by high Pn rates and high P/R ratios during sunny days, low P/R ratios during cloudy days, and low chlorophyll A content. Increased Rni in advanced life cycle stages may reflect an increased ratio of structural to photosynthetic tissue and/or the formation of reproductive organs, both of which require a large amount of energy (LUGO, 1969).

9.4.2 ECOSYSTEM METABOLISM

Large Plexiglas chambers were used to enclose the island ecosystems. The same test system and the same criteria used in the species studies were used to analyze gas exchange within the entire ecosystem.

A water budget was constructed for a favorable summer day (Figure 9). The budget includes water holding capacity of the soil (30387 g/m^2), water stored in plant tissues (781 g/m^2), and water loss through evapo-transpiration (2814 g/m^2). The potential evapo-transpiration was calculated to be 4760 g/m^2, for a ratio of 1.7 which suggests some control of water loss by the plants. At this rate of water loss, soil moisture would be depleted in 10 days.

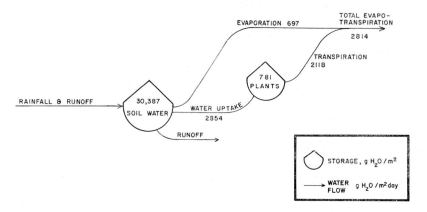

WATER BUDGET FOR A
GRANITE OUTCROP ECOSYSTEM

Fig. 9. The water budget for an outcrop island ecosystem during a day in June. Numbers inside the storage symbols represent the water content of the compartment in g H_2O/m^2. The lines represent the flow of water between compartments and the numbers above the lines represent the rate of water flow in g H_2O/m^2/day. Arrows indicate direction of flow. The diagram is based on data obtained in the North Carolina Botanical Garden. Symbols after ODUM (1967).

This value is consistent with values of 3 days ± 1 for shallow soil and 9 days ± 2 for deep soil reported by CUMMING (1969) for these same ecosystems.

Gas exchange data and determination of calorific values (Table 5) of various components of the ecosystem were used to construct an energy budget for the ecosystem during a favorable summer day (Figure 10). Of the incoming solar radiation, 5.4 % was used in gross photosynthesis. On a single favorable day, the efficiency of net photosynthesis was 2.79 % as compared to 0.32 % for the entire growing season. Based upon visible radiation, the efficiency of Pn is 5.94 % on a favorable day and 0.76 % for the growing season (LUGO, 1969). This value is similar to that reported by BLISS (1960) for arctic and alpine ecosystems. Plants respired 92 % of their production. The P/R ratio of the ecosystem was 1.1 if respiration of soil organic matter from previous years production is not included. If the respiration of previous years production is included, the P/R ratio drops to 0.66.

TABLE 5
Calorific values of selected outcrop species

Species	Calorific value cal/g	Range
Whole plant determinations:		
Polytrichum commune	3892.4	± 9
Juncus georgianus	3991.9	±25
Krigia virginica	3768.0	—*
Allium cuthbertii	3767.9	—*
Selaginella rupestris	3803.6	—*
Senecio tomentosus	3824.6	± 3
Average	3841.4	
Viguiera porteri	4368.5	±20
Cladonia rangiferina	4068.9	± 5
Hypericum gentianoides	4708.1	±16
Panicum sp.	4253.3	±23
Crotonopsis elliptica	4616.8	±12
Portulaca smallii	4133.8	± 5
Bulbostylis capillaris	4191.8	—*
Average	4334.4	
Average of all whole plant determinations:	4006.8	
Plant part determinations:		
Viguiera porteri		
Stem	4581.3	±18
Roots	4534.2	±37
Leaves	4076.1	±19
Inflorescence	5133.1	±13
Standing dead stem	4623.7	± 9
Senecio tomentosus		
Petioles	4223	±16
Roots	4455.8	± 7
Leaves	4760.3	± 9
Inflorescence	5776.5	—*
Attached dead leaves	4943.1	±15

* Denotes only one determination.

9.5 Relationships of Ecosystem Structure and Function

A discussion of structural and functional relationships is obviously a discussion of adaptation. Characteristics of ecosystem structure which are useful in this context are the vertical and horizontal distribution of biomass, energy, and chlorophyll within the system. Similarly, it is useful to compare species morphology,

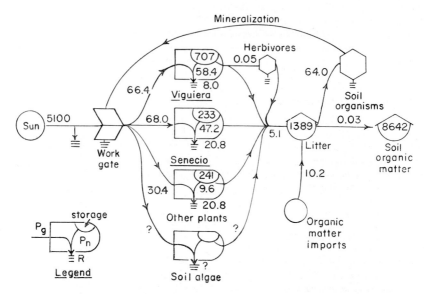

Figure 10. A simplified energy budget for an herbaceous annual-perennial outcrop ecosystem during a period of fast growth (May–June). Numbers inside the storage symbols represent the accumulation of energy in the compartment in Kcal/m². The lines represent the flow of energy between compartments and the numbers above the lines represent the rate of energy flow in Kcal/m²/day. Arrows indicate the direction of flow. The diagram is based on data obtained in the North Carolina Botanical Garden. Respiration is R, net photosynthesis is Pn, and gross photosynthesis is Pg. Symbols after ODUM (1967).

calorific values and chlorophyll content to metabolic patterns and environmental conditions when assessing adaptive strategies.

Species may adapt to severe environmental stress by resisting limiting conditions or by avoiding them. The most severely limiting factor in these ecosystems is water. Accordingly, any method of minimizing transpiration would be of adaptive significance. Transpiration is reduced by extended periods of temporary wilting (*Viguiera*), midday depressions in transpiration, photosynthesis and respiration (*Sedum, Talinum*), inward folding of shoots (*Polytrichum*), and the grassy habit of *Bulbostylis capillaris* (L.) C. B. Clarke. Some species (*Sedum, Minuartia*) avoid severe stress by completing their growth and reproduction prior to the driest seasons.

The group of summer annuals which exhibit explosive growth rates during favorable conditions also exhibit the highest calorific values (4334 cal/g). These species apparently utilize high calorific storages during unfavorable conditions and their survival depends upon a balance of high calorific storages produced during favorable conditions and utilization of stored energy during hot dry periods.

The non-vascular and perennial vascular plants with underground storage organs have low calorific values (3841 cal/g). This group also has lower respiration rates and accordingly, demands upon stored energy are lower. A summary of species calorific values is presented in Table 5.

Chlorophyll A content ranged from 13×10^{-6} g/g dry wet in *Viguiera* stems to 9×10^{-3} g/g dry wt in *Talinum* stems. These relatively low chlorophyll A contents and high Margalef ratios (LUGO, 1969) reflect adaptations to high light intensities characteristic of the rock outcrop habitat.

Total biomass of a typical "island" ecosystem with adequate moisture supply throughout the growing system was 1372 g/m². Dead organic matter consistently comprised over 75 % of the total. In an experimental ecosystem, 354 g/m² was living tissue. Of this, 46 % was in stems, 27 % was in leaves, 2 % was in inflorescences, 13 % was in roots, and 12 % was in dead matter still attached to living plant structures.

Low biomass content in an ecosystem is usually associated with low chlorophyll content (ODUM, MCCONNEL, ABBOTT, 1958), the latter reflecting adaptation to high light intensities. The "island" ecosystem chlorophyll A value of 0.881 g/m² is comparable to that of other light-adapted herbaceous ecosystems (BRAY, 1960a, 1960b) and to arctic and alpine systems (TIESZEN & JOHNSON, 1968).

Experimental analysis has shown that the structure and metabolism of these "island" ecosystems are strongly controlled by microenvironmental conditions, primarily the influence of limited soil moisture and high light intensities upon metabolic rates and the influence of soil depth and moisture upon interspecific competition.

Although photosynthetic rates are relatively high, net production and standing biomass are low due to high respiration rates. High transpiration rates quickly reduce available moisture and consequently rates and efficiencies of production are reduced.

Annual vascular plant species exhibit explosive growth rates under favorable environmental conditions but rapidly deplete high calorific storages during unfavorable conditions. Survival hinges upon the number (approximately 100) and frequency (never less than one in fourteen) of favorable growing days. Non-vascular and perennial species exhibit lower metabolic rates and possess well-developed mechanisms of resisting stress, via dormancy, underground perennating organs, and the folding in of shoots.

Once field and laboratory experiments demonstrated that water availability is a limiting factor which influences ecosystem structure and metabolism the question arose: "What limits population growth and survival during those 100 days per year when

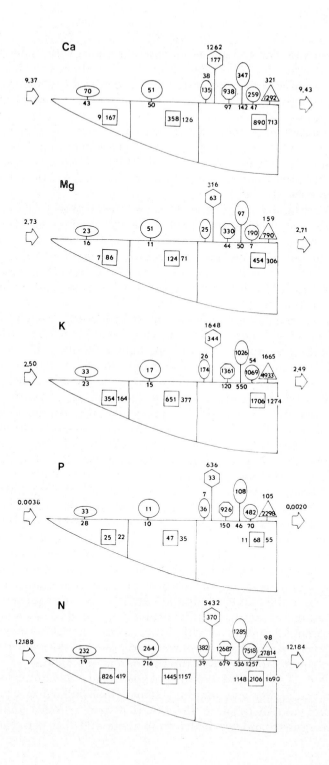

moisture conditions are favorable?" Of the three major factors which most often limit plant growth and reproduction we have eliminated water, and sunlight could hardly be limiting on the outcrops; therefore studies of the influence of nutrient availability were initiated. Extensive analysis of field samples collected from the soil, soil solution, dead organic matter, living biomass, rainfall and rock substrate provided a quantitative description of nutrient distribution and exchanges (Figure 11) in these ecosystems. Comparisons of nutrient uptake and storages by dominant species with concentrations of nutrients available to these species identified what nutrient experiments should be designed for laboratory studies (MEYER, 1972). These comparisons indicated that neither *Sedum* nor *Minuartia* should be limited by nutrient availability in any of the three habitats but that *Viguiera* should be limited by calcium and nitrogen deficiencies in habitats other than its own. However, laboratory experiments (MEYER, 1972) revealed that *Sedum* grows poorly under *Minuartia* habitat conditions due to an inability to compete with *Minuartia* for available nitrogen. *Minuartia* grew poorly under *Viguiera* habitat conditions because of its inability to compete with *Viguiera* for available nitrogen. Both *Minuartia* and *Sedum* obtained sufficient nitrogen and grew successfully in habitats other than their own in the absence of competitors. Levels of available nitrogen and calcium in the *Sedum* and *Minuartia* ha-

---

Figure 11. Nutrient distribution in an 11.6 m² island ecosystem during a 10 week period. Cross sections through part of a typical island ecosystem indicate nutrient content (mg) on June 23rd (date enclosed within the symbols) for plant populations and for the soil solution of three dominant vegetation zones. Species populations are represented by:

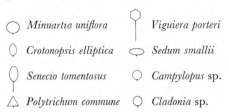

Nutrient uptake per species population is indicated by the data located above plant symbols, while nutrient losses are represented by the data located below each symbol. Initial soil solution nutrient values (mg) are shown in the cubes inset in each of the three dominant species zones. Increases in soil solution nutrients during the time course are indicated to the left of each cube, while decreases are shown to the right. Input of nutrients (g) based upon 12.64 inches of rain during the period are denoted to the left side of the schematic; in contrast, groundwater nutrient output is shown to the right side. Each subsection represents the cycling for one nutrient element.

bitats are insufficient to meet requirements of *Viguiera* populations for these nutrients during the growing season. Therefore it is postulated that under otherwise favorable growing conditions which occur at least 100 days per year, nutrient availability influences population growth, distribution, and thereby influences structure and function of this ecosystem.

Because most of the dominant species are endemic to these ecosystems and because the well-isolated ecosystems exist in a high stress environment, the possibility of species extinction is relatively high. Adaptations to this threat include mechanisms which retain seed within the system over successive seasons and the production of a seed population which has varying fractions which germinate one, two, or more years after dehiscence.

In summary, results of these studies provide a basis for interpreting structural and functional relationships in natural ecosystems. The experimental design outlined in this chapter is intended to serve as one model for the experimental analysis of ecosystems. Results of some of our studies of these manipulatable terrestrial ecosystems were reviewed in an effort to provide examples of the kinds of questions which might be answered by such an experimental approach to ecosystem analysis.

ACKNOWLEDGMENTS

This chapter summarizes research conducted at the University of North Carolina during the past five years. Original and detailed descriptions are available in dissertation form in the library of the Botany Department. We especially wish to acknowledge the work of CLAIR MELLINGER, FAIRMAN CUMMING, KENNETH MEYER and PETER MURPHY. ROBERT PLATT and MADELINE BURBANCK of Emory University first identified the feasibility of using outcrop ecosystems as experimental manipulatable ecosystems. The U.S. Atomic Energy Commission, the National Institutes of Health and The University of North Carolina provided financial support for much of this research.

REFERENCES

BLISS, L. C. – 1962 – Net primary production of tundra ecosystems. pp. 35–44. In Die Stoffproduktion der Pflanzendecke. H. LIETH, (ed.), 156 p., Fischer Verlag, Stuttgart.

BRAY, J. R. – 1960a – The chlorophyll content of some managed plant communities in central Minnesota. *Canad. J. Bot.* 38: 313–333.

Bray, J. R. – 1960b – The primary productivity of vegetation in central Minnesota and its relationship to chlorophyll content and albedo. pp. 102–106. In Die Stoffproduktion der Pflanzendecke. H. Lieth, (ed.).

Burbanck, M. P. & R. B. Platt – 1964 – Granite outcrop communities of the piedmont plateau in Georgia. *Ecology* 45(2): 192–306.

Cumming, F. – 1969 – An experimental design for the analysis of community structure. M.S. thesis, University of North Carolina Botany Department Library, Chapel Hill.

Lugo, A. E. – 1969 – Energy, water and carbon budgets of a granite outcrop community. Ph. D. dissertation, University of North Carolina Botany Department Library, Chapel Hill.

McCormick, J. F. – 1959 – Ecological analysis of selected granite outcrop communities and their response to chronic gamma irradiation. M.S. thesis, Emory University, Atlanta, Georgia. 87 pp.

McCormick, J. F. & R. B. Platt – 1962 – Effects of ionizing radiation on a natural plant community. *Rad. Bot.* 2: 161–188.

McVaugh, R. – 1943 – The vegetation of the granite flatrocks of the southeastern United States. *Ecol. Monog.* 13: 119–166.

Mellinger, C. – 1972 – Ecological life cycle of *Viguiera porteri* and factors responsible for its endemism to granite outcrops of Georgia and Alabama. Ph.D. dissertation, University of North Carolina Botany Department Library, Chapel Hill.

Meyer, K. A. – 1972 – The influence of nutrient availability and nutrient cycles upon ecosystem structure and stability. Ph.D. dissertation. University of North Carolina, Botany Department Library, Chapel Hill.

Murphy, P. G. & J. F. McCormick – 1971 – Ecological effects of acute beta radiation for simulated fallout particles on a natural plant community. pp. 454–481. In: Survival of Food Crops and Livestock in the Event of Nuclear War. U.S. A.E.C. Symposium Series 24, D. W. Bensen & A. H. Sparrow, (eds.) 745 pp.

Odum, H. T. – 1967 – Energetics of world food production. Chapter 3, pp. 55–94. In: The World Food Problem, Vol. III. Report of the President's Science Advisory Committee, Panel on World Food Supply, White House.

Odum, H. T. – 1970 – A Tropical Rain Forest. A Study of Irradiation and Ecology at El Verde, Puerto Rico. U.S. Atomic Energy Commission.

Odum, H. T., W. McConnell & W. Abbott – 1958 – The chlorophyll A of communities. *Publ. Inst. Mar. Sci.* 5: 65–96.

Platt, R. B. & J. F. Griffiths – 1964 – Environmental Measurement and Interpretation. Reinhold, New York.

Platt, R. B. & J. F. McCormick – 1964 – Manipulatable terrestrial ecosystems. *Ecology* 45(3): 649–650.

Sharitz, R. R. & J. F. McCormick – 1973 – Population dynamics of two competing annual plant species. *Ecology* 54: 723–740.

Tieszen, L. L. & P. L. Johnson – 1968 – Pigment structure of some arctic tundra communities. *Ecology* 49(2): 370–373.

Wiggs, D. N. & R. B. Platt – 1962 – Ecology of *Diamorpha cymosa*. *Ecology* 43(4): 654–670.

AUTHOR INDEX

Numbers in boldface type indicate pages on which complete citations to publications are given.

A

ABBOTT, W., 175, **179**
AL-ANI, H. A., 21, **32**
ALLEE, W. C., 25, **32**
ALLEN, L. H., 140, **146**
ANTONOVICS, J., 22, 24, **32**
ARON, W. I., 31, **33**
ASHBY, W. R., 126, **145**
AUCLAIR, A., 101
AUSTIN, M. P., 90, **107**
AUSTIN, R. B., 67, **68**, **69**
AXELROD, D., 116, **121**
AYYAD, M. A. G., 91, **107**

B

BAKER, H. G., 114, **121**
BEALS, E. W., 94, **107**
BEARD, J. S., 116, **121**
VON BERTALANFFY, L., 126, **145**
BECKING, R. W., 77, **84**
BEWS, J. W., 115, **121**
BILLINGS, W. D., 3, 20, 21, 24, 26, **33**, 34, 50, 51, 53, 55, **68**, 89, 96, 97, 99, 102, 103, **107**
BJORKMAN, O., 20, **33**
BLISS, L. C., 172, **178**
BÖRNER, H., 74, **84**
BRADLEY, E. R., **145**
BRADSHAW, A. D., 24, **33**
BRAY, J. R., 56, **68**, 90, 101, **107**, 175, **178**
BREISCH, A. R., 101, **107**
BROUGHAM, R. W., 67, **69**
BROWN, K. W., 137, **147**
BUELL, M. F., 89, **108**
BURBANCK, M. P., 153, 155, 171, 178, **179**
BUSINGER, J. A., 137, **145**

C

CAIN, S. A., 19, 26, **33**, 114, **121**
CANTLON, J. E., 106, **108**
CARMER, S. G., 67, **69**
CHABOT, B. F., 23, **33**
CHASE, V. C., 23, **34-35**
CHOU, C.-H., 79, **84**
CHRISTOPHERSON, J., 24, **33**
CLARKSON, D. T., 22, **33**
CLAUSEN, J., 116, **121**
COLE, N., 116, **121**

COTTAM, G., 73, **84**, 90, 91, **108**
CRICK, H. C., 15, **35**
CRONIN, E. H., 27, **33**
CUMMING, F., 166, 172, 178, **179**
CUNNINGHAM, G., 115, **121**
CURTIS, J. T., 56, **68**, 73, **84**, 90, 91, 100, 101, **108**

D

DALE, M. B., 67, **69**, 90, **108**
DANSEREAU, P., 89, 104, 106, **108**
DARWIN, C., 25
DAUBENMIRE, R. F., 89, 105, **108**
DEBELL, D. S., 76, **84**
DEL MORAL, R., 83, **84**
DICE, L. R., 40, **45**
DIX, R. L., 91, **107**
DUNCAN, P. R., **146**
DUNCAN, W. G., 134, 135, **145**
DUNN, E. L., 116, 117, **122**

E

EDWARDS, N. T., 140, **145**
ELTON, C. S., 39, **45**

F

FERRARI, T. J., 63, **69**
FOSTER, R. B., 30, **33**
FRIEDERICKS, K., 25, **33**
FURNIVAL, G. V., 140, **147**

G

GAASTRA, P., 129, **145**
GATES, D. M., 95, 97, **108**, 109, 115, **121**, 129, 131, **146**
GAY, L.W., 131, **146**
GEIER, M. R., **34**
GEIGER, R., 103, **109**
GIMINGHAM, C. H., 104, **108**
GLEASON, H. A., 93, **108**
GLENDAY, A. C., 67, **69**
GLIESSMAN, S. R., 82, **84**
GOFF, F. G., 90, 91, **108**
GOOD, R., 19, **33**
GREIG-SMITH, P., 91, **108**
GRIFFITHS, J. F., 103, **109**, 155, **179**
GRINNELL, J., 39, 40, **45**

H

HAASE, E. F., 105, **108**
HALL, A. E., 127, **146**
HANAU, R., **145**

HANAWALT, R. B., 77, **85**
HARPER, J. L., 73, **84**
HARRISON, A. T., 23, **34**
HESLOP-HARRISON, J., 20, **34**, 115, **121**
HOLMGREN, P., 20, **33**
HOLWAY, G. G., 99, 102, **109**
HOOK, D. D., 74, **84**
HUTCHINSON, G. E., 40, **45**

I

IZUMI, Y., **145**

J

JACCARD, P., 56, **69**
JACOBS, J. A., 67, **69**
JOHNSON, P. L., 175, **179**

K

KEELING, D., 32
KNOERR, K. R., 5, 131, 140, **146**
KORNBERG, 15
KREITH, F., 136, **146**
KRUCKEBERG, A. R., 21, 22, **34**

L

LAMBERT, J. M., 90, **108**
LANCE, G. N., 56, **69**
LANGENHEIM, J., 10, **34**, 50, **69**, 95, 96, **108**
LANGFORD, A. N., 89, **108**
LEDERBERG, 15
LEMON, E. R., 129, 140, **146**
LEVINS, R., 44, **45**
LEWONTIN, R. C., 45, **45**
LIEBIG, 17
LIETH, H., 106, 107, **108**
LIVINGSTONE, B. E., 114, **121**
LOACH, K., 140, **146**
LOMMEN, R. W., 127, **146**
LOOMIS, R. S., **145**
LOUCKS, O. L., 91, 97, 103, 105, **108**
LUGO, A., 5, 155, 160, 167, 170, 171, 175, **179**

Mc

MCCONNEL, W., 175, **179**
MCCORMICK, J. F., 5, 155, 160, 166, 167, **179**
MCINTOSH, R. P., 89, 90, 91, 100, 101, 104, 106, 107, **108**
MCKELL, C., 21, **34**
MCMILLAN, C., 20, **34**
MCPHERSON, J. K., 77, 78, 79, **85**

MCVAUGH, R., 152, **179**
MACHTA, L., 32, **34**

M

MAGUIRE, B., JR., 41, **45**
MAJOR, J., 50, **69**, 91, 96, 103, **109**, 113, **121**
MARR, J. W., 105, **108**
MARTIN, F. F., 126, **146**
MASON, H. L., 9, 19, 22, **34**, 50, 69, 95, 96, **108**
MAXIMOV, N. A., 115, **121**
MAYCOCK, P. F., 91, **108**
MELLINGER, C., 155, 164, 178, **179**
MERRIL, C. R., 15, **34**
MEYER, K. A., 177, 178, **179**
MOEN, A. M., 95, **108**
MOLISCH, H., 73, **84**
MONK, C. D., 91, 97, **108–109**
MOONEY, H. A., 4, 20, 23, **32**, **34**, 115, 116, 117, **121**, **122**
MUIR, JOHN, 25, **34**
MULLER, C. H., 4, 73, 77–83, **84**, **85**
MURPHY, C. E., JR., 5, 140, **146**
MURPHY, P. G., 167, 178, **179**
MURRAY, F. W., 133, **146**

N

NELDER, J. A., 67, 68, **69**
NIERING, W. A., 91, 105, **109**, 117, **122**

O

ODUM, E. P., 15, **34**, 106, **109**
ODUM, H. T., 17, **34**, 154, 175, **179**
OLMSTEAD, C. E., 20, **34**, 81, **85**
OLSON, J., 32
OOSTING, H. J., 74, **85**
ORLOCI, L., 90, **107**

P

PALTRIDGE, G. W., 141, **146**
PARENTI, R. L., 81, **85**
PARKS, T., 25, **32**
PARKHURST, D. F., 129, **146**
PATTEN, B. C., 126, **146**
PATTEN, D., 27, **34**
PERRY, T. O., 20, **34**
PETRICIANNI, J. C., **34**
PLATT, R. B., 5, 103, **109**, 152, 154, 155, 160, 166, 171, 178, **179**
PORTER, W. P., 95, **109**
PROSSER, C. L., 23, **34**
PUTTER, J., 64, **69**

R

RABINOWITCH, E. I., 127, **146**
RADEMACHER, B., 74, **84**
RANDALL, J. M., 56, **69**
RASCHKE, K., 131, **146**
RAUNKIAER, C., 114, **122**
REIFSNYDER, W. E., 140, **147**
RICE, E. L., 81, **85**
RUNE, O., 21, **34**

S

SALISBURY, F. B., 53, **69**
SAMPSON, A. W., 79, **85**
SCHWINTZER, C. R., **146**
SCOTT, D., 4, 20, **34**, 60, 63, **69**, 89, 96
SCOTT, J. T., 4, 57, 99, 102, **109**
SEMIKHATOVA, O. A., 23, **34**
SHARITZ, R. R., 155, 160, 167, **179**
SHAWCROFT, R. W., 129, **146**
SHREVE, F., 81, 85, 114, 117, **121**, **122**
SICCAMA, T. G., 102, **109**
SINCLAIR, T. R., 5, 127, 140, **146**
SMITH, S. H., 31, **33**
SNAYDON, 24
SOLLINS, P., 140, **145**
STEWART, D. W., 127, 129, 140, **146**
STRAIN, B. R., 23, **32**, **34-35**, 115, **121**
STUBBS, J., 74, **84**
SWEENEY, J. R., 79, **85**

T

TANSLEY, A. G., 22, **35**
TEERI, J. A., 24, **35**
THOM, A. S., 129, **146**
TIESZEN, L. L., 175, **179**
TOBIESSEN, P., 21, **35**

TUKEY, J. W., 63, **69**
TRACY, J. G., 56, **69**
TURESSON, G., 19, **35**, 116, **122**
TURNER, N. C., 140, **147**

U

UCHIJIMA, A., 137, **146**
UDVANDY, M., 39, **45**

V

VAARTAJA, O., 20, **35**
VISSER, W. C., 65, 67, **69**

W

WAGGONER, P. E., 127, 130, 140, **146-147**
WARING, R. H., 91, 103, **109**
WATSON, J. D., 15, **35**
WATT, K. E. F., 45, **45**, 65, **69**
WEATHERLEY, A. H., 39, **45**
WEBB, L. J., 116, **122**
WEST, M., 23, **34**
WHITTAKER, R. H., 4, 41, 44, **45**, 90, 91, 93, 94, 96, 105, 107, **109**, 117, **122**
WIGGS, D. N., 155, 160, **179**
WILLIAMS, W. A., **145**
WILLIAMS, W. T., 56, 67, **69**
WILSON, R. E., 81, **85**
WOODWELL, G., 32
WOODS, D. B., 140, **147**
WRIGHT, J. L., 137, **146**, **147**
WUENSCHER, J. E., 3, 42, **45**, 115, **122**
WYNGAARD, J. C., **145**

Y

YOCUM, C. S., **146**

Z

ZINDER, 15

SUBJECT INDEX

No attempt has been made to index every occurrence of all terms. Repeating terms are indexed only to the most definitive or important occurrences.

A

Abies balsamea, 100
    fir seedlings, 101
Abundance, 155, 167
Acclimation, 23–24
Acclimatization, 19, 23–24
Acer rubrum, 20
    saccharum, 78, 100
Achillea borealis, 116
    lanulosa, 116
Adaptation, 4, 19, 21, 22, 117, 160, 171, 173, 177
Adenostoma fasciculatum, 77–79, 82
Africa, 4
Agropyron sp., 27
Agrostis canina, 24
    setacea, 22
Alabama, 152
Alaska, 29
Allelopathy, 4
    definition, 73–74
    exclusion, 77–78
    function of in dominance, 76–79, 82
    in abandoned fields, 81–82
    limits of, 82–84
    measurement of, 74
    phytotoxic, 73
    relation to competition, 74–76
    synergisms, 73
    toxic elements, 51, 52, 73
Allium cuthbertii, 173
Allopatry, 76
Alluvial fan, 21
Alosa pseudoharengus (alewife), 31
Alpine,
    plants 20, 23, 24, 172
    region 21
Altitude, 28, 55, 91, 94, 96, 99–102
Anatomy,
    ecological, 114–115
Arctic, 14, 21
    plants, 23, 24, 172
    tundra, response to disturbance, 81
Arctostaphylos glandulosa var. *zacaensis*, 77, 80
    sp., 82
Arizona, 105
Aromatic species, 83

Artemisia sp., 27
Aspect, 28, 51, 55, 91, 99–101
Association, plant, 4, 91, 101, 105, 106
Atmosphere, 12, 14
    bouudary layer, 126
    toxins in, 83
Atmospheric gases, 12, 51
    cycles, 16
Autecology, 3
Autointoxication, 52, 73, 82
Auto-succession, 81

B

Bacteria, 16
Barro Colorado Island, 31
Beetle, pine, 29
Behavior, 4
    and form, 117, 119
    correlations with other factors, 159
Betula allaghaniensis, 100
    papyrifera var. *cordifolia*, 100
Bioassay, 79, 83
Biochemistry, 4
    role of in vegetation stability, 81–82
Biochemical, 73, 74
    adaptation, 115
    control of herb growth, 80
    role of in vegetation stability, 81
Biological accumulation, 29
Biomass, 82, 153, 155, 173
    and competition, 161–162
    and soil depth, 159, 162
    of rock outcrop ecosystem, 175
    pyramid, 15–16
Biosphere, 14, 73, 83
Biota, 119
Bitterroot Mountains, 105
Bouyoucos blocks, 166
Bromus tectorum, 27–28
Bulbostylis capillaris, 168, 173, 174

C

Cactaceae, 4
Calcium, 14, 22, 28, 55, 177
California, 14, 23, 25, 29, 77, 105, 116
Calorific values, 172–175
Campylopus sp., 177
Canal,
    Erie, 31
    Panama, 31
    Suez, 31
    Welland, 31

Canopy, 126, 135–138
Carbon, 4
    balance, 116, 120
    fixation, 125
Carbon dioxide, 14, 19, 51, 125
    atmospheric concentration, 32, 130, 155
    exchange rates, 170–171
    fixation, 126
    use in modeling, 127–130, 140–143
    vertical gradients of, 136–138
Carnivore, 16, 29
Celmisia spectabilis, 55
Chaparral, 77
Chile, 116
Chionochloa rubra, 55
Chlorophyll, 171, 173, 175
Chloroplasts, 125
Chromatography,
    pigment separation, 166
Cladonia rangiferina, 173
    sp., 177
Classification,
    agglomerative method, 56
    devisive method, 56
    phylogenetic, 114
    similarity coefficients, 56
    vegetation, 4, 90, 115
Clay, 28, 51, 55
    and phytotoxicity, 83
Climate, 4, 19, 51, 53, 90, 96, 102
    and growth form, 114
    ecoclines, 22
    ecotypes, 22
    homologous, 114, 119
    macro-, 113
    Mediterranean, 4, 113, 117–119
    micro-, 127, 130
    models, 133–134
Climax concept, 81
Clone, 23
Coastal, 21, 23, 116
Cohort, 161
Colonization, 13
Colorado Front Range, 105
Community,
    aquatic, 12
    concept, 44, 89–107
    continuity, 44, 89, 93–94, 106
    destruction of, 13
    discontinuity, 44, 93–94, 106
    environment, 4, 10, 12–13
    hypotheses, 89, 92–93, 106–107
    index, 103
    monitoring of, 32
    plant, 43, 105, 113
    ordination of, 41, 92–93
    requirements for sampling, 91
    stability, 43
    succession, 13, 43
    terrestrial, 43
Compensation, in ecosystems, 151
Competition, 4, 14, 22, 40, 42, 52, 106
    and growth form, 119
    effect on community structure, 94
    interspecific, 101, 160, 161, 167, 175
    intraspecific, 160–164
    on granite outcrops, 155, 164, 177
    relation to allelopathy, 74–76
Competitive exclusion, 50, 78
    effect on community structure, 94
    in prairie vegetation, 82
    relation to allelopathy, 76
Computer, 27, 32, 68, 125, 126, 133, 143
Conduction, 13
Consumers, 120, 121
Continuum,
    analysis, 107
    concept, 44, 89
    community continuity, 44
    genetic, 22
    hypothesis, 89, 92–93, 101, 105, 106–107
    requirements for sampling, 91, 103
    space-time, 12
Convection, 127
Convergence, 4, 113–114, 119–121
Coreopsis grandiflora, 169
Coumarin, 79
Cover, 52
Crotonopsis elliptica, 167, 171, 173, 177
Cyperus granitophilus, 167

D

Dactylis glomerata, 21
Dansereau's Concert, 104, 106
DDT, 29
Decomposer, 39
Decomposition,
    in granite outcrops, 152
    in relation to phytotoxicity, 77
    in relation to vegetation stability, 81
Density, 153, 155, 157
    and competition, 161–162
    dependent, 158
    independent, 158
Desert,
    cold, 78

environment, 9
evolutionary convergence in, 119
lack of allelopathy in, 83
plants, 21, 23, 115
polar, 14
response to disturbance, 81
Detritus, 15–16
Devon Island, 24
Diagram,
causal, 4, 54, 55, 62, 68
structural, 54
Dispersal, 52, 114
Distribution, 155, 167
Disturbance,
and homeostasis, 81
effect on vegetation, 29–30, 89, 97
revegetation following, 81, 97
Diversity, 43–44, 106
DNA as an environmental component, 15–19
Dominance, 4
and allelopathy, 76–79
and growth form, 113–114, 117
and stability, 81–82
functional definition, 76
in forests, 74
in relation to phytotoxicity, 74
mixed, 94
role of biochemistry in, 81–82
single species, 78–79
time dimension, 77
stand, 100
Dracophyllum recurvum, 55
Drought, 18, 21
evaders, 117, 119
relation to phytotoxicity, 77
resistance, 117
tolerance, 117
Dryas sp., 14

E

Ecesis, 81
Ecocline, 21, 22
Ecosystem,
allelopathy, 82
analysis of, 151–178
behavior, 4
boundaries, 152
change, 15, 32
closed, 152
concept, 151–152
control of, 16
definition of, 9
disturbance of, 31

energy budget, 174
experimental analysis of, 154–172
function of, 15–16, 154, 167–178
holocoen, 25
holocoenotic nature of, 25–32
homeostasis, 81–82
"island", 151–154, 164, 166, 169, 175–177
man dominance of, 15
metabolism, 171–172
monitoring of, 32
natural, 5, 15, 163
structure, 4, 154–167, 173–178
theory, 3
water budget, 172
Ecotone, 93, 94, 101
Ecotype, 19–22, 24, 94, 116
Edaphic (see soil)
Eddy size, 137
Elevation, (see altitude)
Emigration, 152
Encelia californica, 23
Endemics, 21–22, 153, 178
Energetics,
of community structure, 106
of plant adaptation, 4
Energy,
balance, 127, 131–133
budget, 170, 174
exchange, 51
flow, 14, 16–17, 131–133
limited supply, 74
of leaves, 127, 131–133
regulation, 120
Environment,
aquatic, 12–13
axes, 40, 43
biological, 12
cellular level, 15
coastal, 9
community level, 96, 97, 103, 106, 107
complex, 9–32, 95–96
desert, 9
dimensions of, 11–13, 41–42
ecosystem structure, 120
factors, 96–102
generalized, 90
global monitoring, 32
gradients, 34, 40, 42, 91–93, 107, 116–117
holism, 3, 4, 16, 25–32, 50, 55, 96, 102
instability, 22

186

linearization, 92, 103–104
lunar, 9
matrix, 42
micro-, 175
molecular level, 15
operational, 9–11, 13, 17
optimal, 163
organism level, 15, 92–93, 97, 103, 107
physical, 12, 16
potential, 10, 107
relation to yield, 66
rural, 9
source domain, 12, 13
space, 12, 13
space-time continuum, 12
species level, 9–10, 13, 17–32
terrestrial, 12, 13
time, 12, 14
total, 11
urban, 9
Environmental conditioning,
of organisms, 23–24, 43, 49–50, 120
of sites, 107
Environmental index,
definition, 90
measure of site potential, 98–99
need for independent variables, 91, 92
use of, 98–99, 105
Environmental Protection Agency, 29
Environmentization, 23
Enzymes, 17
reaction mediation, 125
ribulose–1, 5-diphosphate carboxylase, 23–24
Eriogonum sp., 116
Erosion, 29–30, 51–52
Eucalyptus camaldulensis, 83
Euphorbiaceae, 4
Evapo-transpiration, 168, 171
Evolution, 106
convergent, 119
of growth form, 113, 115
parallel, 116
Extinction of species, 15, 178
Exudation, 81
Evaporation, 51, 129–132
Exfoliation, 81

F

Fagus grandifolia, 78, 100
Feedback, 17, 18
Fertilizer, 23

Festuca ovina, 24
Fire, 12, 14, 15, 27, 51, 53, 117
relation to phytotoxicity 77, 78–79
Fitting of functions,
criteria for, 65
empirical equations, 57–58, 65
extrapolation, 58
in ordination, 92, 97, 101
iterative techniques, 65, 68
seven most common equations, 65
stages in, 57, 59–60
testing of significance, 60
theoretical equations, 57–58, 65
Florida, 20
Food chain, 14–15
Food web, 14–15
Forest,
climax, 81
coniferous, 14, 78
mixed, 74, 76
montane tropical, 82
tropical rain, 154
upland, 105
vegetation, 99
Form,
and function, 114–116
behavioral, 4, 117, 119
plant in relation to environment, 114–117, 118
RAUNKIAER classification, 114
Frequency, 55, 153
Fruits, 30–31, 52
toxins in, 83
Function, 5
ecosystem, 167–178
Fungi, 16

G

Galium saxatile, 22
sylvestre, 22
Gases (see atmospheric gases)
Genecology, 19–20, 115
Genotype, 10, 22
Geography, plant, 4, 53, 114
Georgia, 152, 157, 159
Germination, 77, 78, 155, 164
Goff-Gratch equation, 133
Gopher, pocket, 28
Gradient analysis, 107
Granite outcrops, 5, 152–178
Grass, 27, 77
Grasslands, 14, 113
climax, 81, 82
Gravel, 28

Gravity, 10, 12–14
Grazing, 15–16, 52, 118, 152
Great Basin, 27, 78
Green Mountains, 102
Growth,
    estimation of, 67–68
    inhibition of, 73, 76, 78
    logistic curve, 66
    of granite outcrop species, 171, 175
    simulation of, 68
    stand, 127
    stimulation of, 73
    toxic suppression of, 73, 74, 77, 78–79
Growth form, 113–114, 117–119
Groundwater, 21

H

Habitat, 4, 42, 43, 78, 81, 153, 177
Halogeton glomeratus, 27–28
Hawaii, 32
Heath, 82
Heavy metals,
    accumulation, 13
    concentration, 21
    sedimentation, 13
    tolerance to, 27
Helianthus, sp., 73
Heliophytes, 170
Herbivore, 16
Herbs, 77, 79, 83, 153
Hieracium umbellatum, 116
Holocoenosis, 3, 4, 107
    diagram, 26, 28
    holocoenotic environment, 25–32, 96
Holism, 16, 50, 55, 102
Homeostasis,
    ideal, 24
    partial, 24
    role of in analyzing ecosystems, 82
Homeostatic vegetation, 77, 81
Humus, 28, 52
Hunting, 15
Hypericum gentianoides, 169, 171, 173
Hyperspace, 3, 13, 40–41, 44

I

Immigration, 152
Importance index, 90, 94–95
Indo–Pacific, 31
Insects, 14
Interference, 4, 74
    biochemical, 76
    definition, 74
    effect on regeneration, 76
    need for research, 78
International Biological Programme, 32, 119, 142
Invasion,
    grasslands by woody plants, 77
    homeostatic resistance to, 81
Interrelationships, 3
Investigator point of view, 53
Isomeris arborea, 21

J

Juglans nigra, 76–77
    sp., 82
Juncus georgianus, 173
Juniperus sp., 82

K

von Karman's constant, 136
Krigia virginica, 173

L

Lake Erie, 31
Lake Ontario, 31
Lake Gatun, 31
Lakeshore, 91
Landscape, 113
Latent heat of vaporization, 132
Latitude, 20, 51
Lava, 51
Leaching, 81
Leaf,
    area and photosynthesis, 135
    energy balance, 131–133
    optical properties, 135
    photosynthesis, 127–131
    temperature, 136
    water potential, 21
Lichen, 153
Life form (see form)
Life tables, 155, 156
Light (see radiation)
Limiting factors, 17–25
    Mitscherlich law of, 66
    equations, 66
    phytotoxicity, 73
Limits of survival, 40
Liquidambar styraciflua, 76
Liriodendron tulipifera, 140–143
Lithosphere, 12, 14
Litter, 28, 51, 52, 77, 79
Longitude, 51

M

Magnesium, 28, 55
Man,
    fire, 15
    hunting, 15
    gathering, 15
    gardening, 15
    impact on global ecosystem, 31, 52
    pollution effects, 29
    technology, 15
MARGALEF ratio, 175
Mauna Loa Observatory, 32
Meadow, wet, 24
Mediterranean, 4, 21, 27, 31, 113, 117, 118, 119
Metabolism,
    of outcrop ecosystem, 171–172
    of outcrop species, 170–171
Meteorology, 125
    vertical profiles, 127
Meteorite, 12
Microorganisms, 12, 15, 79
Migration,
    man made canals and, 31
    effect on populations, 31
Mineral cycling (see nutrient cycling)
Minerals, 12, 14, 51–52, 74, 97
    conservation of, 117
    in rock outcrop ecosystem, 176–177
MINKOWSKI-EINSTEIN,
    curved space-time, 14
Minuartia uniflora, 153, 155, 157–160, 162–164, 167
Modeling
    carbon dioxide, 32
    curve fitting, 57–59
    empirical, 65, 68
    evolutionary, 118
    hypothetical, 27
    input, 130
    leaf photosynthesis, 127–131
    mathematical, 5, 125–142
    microclimate, 130
    non-steady state, 138
    objectives, 126
    photosynthesis of plant stands, 126, 138–140, 140–143
    predictive, 32, 42, 58, 63, 126, 140
    simulation, 58, 68, 130, 138, 141
    stand growth, 127
    stochastic, 92, 97
    stomatal, 129
    sub-process, 5, 127
    theoretical 65, 68
    validation, 5, 127
    whole system, 5
Monkeys, 31
Mortality, homeostatic resistance to, 81
Mountain plants, 23
Mt. Arabia, 152, 155, 157, 159

N

N-dimensional (see niche)
New Brunswick, 105
New York, 90–100
Niche, ecological
    change of, 42–43
    competitively realized, 42
    concept, 39–45
    definition, 39–40
    differentiation, 5
    experimental analysis of, 41
    n-dimensional, 3, 4, 40
        hyperspace, 40, 41
        hypervolume, 40, 41, 42, 43
    overlap, 42
    parameters, 3
    potential, 40, 42
    realized, 40, 42
    response vectors, 42
    space, 42, 43, 44
    specialization, 117–119
    species response volume, 41
    succession, 43
    vector space, 41
Nitrifying organisms, 82
Nitrogen, 18, 28, 81
    competition, 177
    fixing organisms, 82
    nitrate, 14
Noda, 106
North America, 14, 20, 21, 27, 119
North Carolina, 152, 165, 172
Notodanthonia setifolia, 55
Nutrient cycling, 16, 51, 52, 82, 106
Nutrients (see minerals)

O

Oklahoma, 81
Ordination,
    Australian forest types, 56
    community index, 103
    compositional index 90
    graphical, 55
    environmental index, 90–93
    linear indices, 90, 92
    multi-dimension, 90

of vegetation, 90
  requirements for sampling, 91, 92
  similarity index, 101
  strip method, 100
  vegetation index, 90, 92-93, 97-98, 100-101
Oregon, 105
Oxygen, 14, 51
  inhibition of photosynthesis, 23
Oxyria digyna, 20, 23, 24
Ozone, 51

P

Paleo-,
  climate, 114
  ecology, 114
Palm, spiney, 31
Panama, 29
Panicum sp., 173
Pathogen, 18
Pattern, 167
Pentromyzon marinus (sea lamprey), 31
Permafrost, 29
Pesticide (see DDT)
Phenols, 78, 79, 83
  phenolic glycosides, 83
Phenotype, 15, 22
Phosphorus, 14, 28, 55
Photoperiod, 20, 53
  dormancy, 20
  ecotypes, 20
  flowering, 20
Photosynthesis, 14, 16, 19, 20
  acclimation, 19
  as a niche parameter, 41
  carbon dioxide, 19, 127-131
  ecosystem 169
  ecotype, 20, 21, 24
  gross, 128, 172, 174
  Hill reaction, 23, 24
  light, 19, 20, 57
  modeling, 125, 127-131, 133
  non-linearity, 134
  of granite outcrop species, 169, 170-171, 174-175
  optimum environment, 19
  stand, 138-140, 140-143
  temperature, 19, 20, 24, 127-131, 131-133
  vapor pressure, 19
  water potential, 21
Physiognomy (see also growth form), 105, 113

Phytotoxicity, 4, 73
  definition, 74
  synergisms, 77
Phytotron, 19
Picea rubens, 100
  spruce seedlings, 101
Pineapple, 23
Pinus attenuata, 14
  ponderosa, 29
  serotina, 14
Pioneer species, 82
Plasticity, phenotypic, 22-23, 24
Poa sp., 27
Polar, 29
Pollen, 14
Pollination, 10
Pollutants, 12, 28, 51
Polytrichum commune, 168, 171, 173, 174, 177
  mats, 164
Pond, 152
Population, 3, 10, 106
  disturbance of, 31, 166-167
Populus grandidentata, 100
Portulaca smallii, 173
Potassium, 28, 55
Prairie,
  homeostasis in, 82
PRANDTL number, 129, 132
Precipitation, 12
  relation to phytotoxicity, 83
Predator, 117
Pressure, 12, 51
Primary producer, 39
Productivity, 82, 120, 153
Provenance, 20
P/R ratio, 170-172
Pteridium aquilinum, 82
Puerto Rico, 154
Pyric (see fire

Q

Quercus falcata var. *pagodaefolia*, 76
  marilandica, 77
  sp., 82
  stellata, 77
  rubra var. *borealis*, 100
Quinones, 79

R

Races,
  coastal bluff, 116
Radiation
  absorptivity, 132-133
  cosmic, 12-13

diffuse, 134
direct beam, 134
effect on plants, 19, 53
electromagnetic, 12
emittance, 131
flux, 127, 131–133
heat, 12
light, 19, 20, 52, 74, 134
longwave, 131–133
periodicity, 53
photosynthetically active 19, 128, 133, 134
shortwave, 131–133
sky, 12, 16
solar, 12, 13, 16, 51, 55, 134, 155
terrestrial, 51
thermal, 12
transmitted, 135
view factors, 135–136
Radioactivity, 12
fallout, 12
radionuclides, 29
Rainforest, 14
Recurrent arrays, 106
Reproduction,
and convergence, 114
and ecosystem structure, 155
effect of aseasonal rain, 30–31
environmental regulation of, 17
fire, 14
potential, 155
Resistance,
boundary layer, 128, 132
diffusion equation, 128
drought, 117
mesophyll, 23
protoplasm, 128
stomatal, 23, 128–129, 132
wall, 128
Resources, biological, 119
Respiration, 14
acclimatization, 24
base rate, 130
community, 16
ecosystem, 169
ecotypes, 21
mitochondrial, 128, 130
niche parameter, 41
of granite outcrop species, 169, 170–171, 174, 175
soil, 169
Rhacomitrium lanuginosum, 55
RIEMANNIAN space-time continuum, 13
RNA, 17–19

Rock, 12, 21, 22, 28, 102, 113
outcrops, 151–154, 165, 166, 175
Root system, 21, 78
Rosette, 153, 156, 157

S

Sagebrush, 27
Salicyclic acid, 76
Salvia leucophylla, 77
Sand, 28
and phytotoxicity, 83
San Francisco Peak, 105
Santa Catalina Mountains, 105
Savannas, tropical, 14
Saxifraga oppositifolia, 24
sp., 14
Scalers, 53, 103
SCHMIDT number, 129, 132
Schoenus pauciflorus, 55
Sedum smallii, 153, 155–163, 167, 174, 177
Seed production, 162, 163
Seedling, 101, 153, 157, 170
Selaginella rupestris, 168, 171, 173
Senecio tomentosus, 154, 167–171, 173, 177
Shrubs, 77, 79, 118
deciduous, 116
evergreen, 116
sclerophyllous, 113, 117
Simulation of photosynthesis, 127, 140–142
Seeds, 30–31, 52
Selection, 116
Shade-tolerant plants, 74
Simmondsia chinensis, 21
Smog, 29
Los Angeles, 29
photochemical, 29
Snowbank, 24
Soil
acid, 152
aeration, 51, 55
bulk density, 28, 55
depth, 155, 162–163
ecotype, 21, 22
evapo-transpiration, 168
field capacity, 28
glacial, 102
limestone, 21
moisture, 28, 51, 55, 74, 155, 171–172
nutrients, 97, 177
organic matter, 28, 51–52, 55
peat, 29
pH, 28, 51, 55, 164

phosphorus deficient, 18
profile, 28, 51, 55, 78
respiration, 152
sandy, 152
serpentine, 21, 113
structure, 28, 51
texture, 28, 51
toxicity, 76, 79, 83
Solidago virgaurea, 20
Sonoran Desert, 21
South America, 119
South Carolina, 76
Space, 89
    as an environmental dimension, 151
    part of living regime, 51
    environmental variation in, 91, 99–102
    universal, 12, 16
Stand, 90, 91, 92, 97–98
    management, 126
    photosynthesis of, 125–127, 138–140
    simulation of photosynthesis, 140–142
Stability,
    community hypotheses, 106
    definition, 43
    related to allelopathy, 81–82
    related to dominance, 81–82
Stand photosynthesis solution scheme, 139
Statistics
    analysis of variance, 52, 60–61, 63, 68
    association analysis, 68
    canonical correlation, 52, 60
    coding, 61
    component analysis, 68
    confidence intervals, 63, 68
    contingency table, 52, 60–61, 68
    correlation, 3, 52, 61, 64
    covariance, 52, 60–61, 63
    curvilinear relationship, 64–68
    dependent variable, 60–62
    discriminant analysis, 52, 60–61
    empirical equations 65, 68
    error, 60–61, 63
    factor analysis, 52, 60–62, 68
    factor variable, 60
    fitting of functions, 57–68
    GAUSSIAN distribution, 61
    independent variable, 60, 61, 63
    least squares deviation, 61
    linear relationship, 59–64, 65
    logarithmic transformation, 64
    multistage relationship, 63–64, 68

    normal distribution, 61
    null hypothesis, 61
    path analysis, 52, 60
    polynomial transformation, 64
    principal components analysis, 52, 60
    probability, 68
    quadratic relationship, 64
    randomness, 60–61, 91–92
    regression, 3, 52, 60–63, 65, 68, 101, 103
    residual effect, 60–61
    response surface, 65
    significance levels, 27, 28, 60, 63
    similarity coefficients, 56
    similarity index, 101
    subject variable, 60
    theoretical equations, 65, 68
    validity of assumptions, 64
STEFAN-BOLTZMANN constant, 132
Steppe, 105
Stomata, 23
    resistance, 127–131
Strategy
    adaptive, 114, 117, 171
    evolutionary, 118
    form-behavioral, 4, 117
Stratification,
    as an ecosystem characteristic, 5
    of seeds, 164
Structure, 4, 5, 167
Sub-polar, 29
Succession, 5, 42–43, 97–99, 106
    and homeostasis, 81
    auto-, 81
    of abandoned fields, 81
    of growth form, 113
    secondary, 82, 97
Succulent, 4, 116
Sugar, concentration in plants, 171
Survivorship, 158
Swamp, 91
Symbiosis, and stability of vegetation, 81
Synecology, 3, 74
Synergism,
    in ecosystems, 151
    with phytotoxins, 77
System,
    biological simulation of, 68, 127
    ecological, 9–10
    intact, 5
    stand photosynthesis, 126
    sub-, 126

T
Talinum teretifolium, 167, 168, 169, 174, 175
Temperature,
  acclimatization, 23
  air, 51, 155
  biosphere, 4, 73
  ecotypes, 20–21
  effect on photosynthesis, 19, 130
  frost, 102
  soil, 51, 155
  transpiration, 115
Terpenes, 83
Thuja occidentalis, 100
Time, 42–43, 51, 53, 82, 89, 97–98
  as an environmental dimension, 151
Tolerance,
  concept, 17–25
  range, 40, 49
  theory of, 19
  to allelopathy, 76
  to shade, 76
  understory species, 77
Topography, 51, 53, 74, 79, 99
Tortuosity factor, 129
Transduction, 15
Transfer processes,
  convective, 134, 136–138
  radiative, 134–136
Transpiration,
  and leaf size, 115
  as a niche parameter, 41
  of granite outcrop species, 168, 174, 175
  of *Senecio tomentosus*, 168, 170
Tree,
  coniferous, 78, 113
  evergreen, 113, 116
  seedlings, 76
Trigger factor, 25, 27, 31
  man imposed, 29
Trophic
  dynamics, 39
  level, 15–16, 29
Tsuga canadensis, 78, 100

Tundra, 29
  response to disturbance, 29–31
Turbulent transfer,
  of heat, 55
  of water vapor, 55

V
Vapor pressure, 19
Vegetation, as a function of environment, 96, 106–107
Vermont, 20
Viguiera porteri, 153–154, 159–160, 163–164, 166–171, 173–175, 177
Vulcanism, 51

W
Water,
  balance, 116–117
  budget, 170–172
  cloud, 12, 51, 102
  cycle, 16, 51
  fog, 83
  loss, 120
  precipitation, 12, 51, 77, 83
  soil, 28, 51, 74, 77
  vapor, 12, 51
Weeds, as pioneers, 82
Wheat, 18
Whiteface Mountain, 99–102
Wind, 12, 51, 55, 129–130
  effect on photosynthesis, 141
Wisconsin, 105
Woodlands, 113, 116

Y
Yellowstone National Park, 27, 28
Yield,
  equations for estimation of, 66
  relation to environmental factors, 66

Z
Zinc, 24
Zonation, 5, 167, 176–177
Zygote, 17